What Is Life ?

1
Earth in space.

Between these two realms

2
Mycoplasma. Phylum: Aphragmabacteria. Kingdom: Monera. Some of the smallest bacteria (magnified 10,000 times).

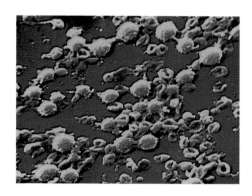

WITHDRAWN

lies all of life as we know it.

What Is Life?

Lynn Margulis and Dorion Sagan

Foreword by Niles Eldredge

A Peter N. Nevraumont Book

SIMON & SCHUSTER

New York London Toronto Sydney Tokyo Singapore

SIMON & SCHUSTER
Rockefeller Center
1230 Avenue of the Americas
New York, New York 10020

10 9 8 7 6 5 4 3 2 1

Printed in Italy

Library of Congress Cataloging-in-Publication Data

Margulis, Lynn, 1938-
    What is life? / Lynn Margulis and Dorion Sagan :
introduction by Niles Eldredge.
        p.   cm.
    "A Peter N. Nevraumont book."
    Includes bibliographical references.
    ISBN 0-684-81326-2
    1. Life (Biology) 2. Biology—Philosophy.
3. Biological diversity. 4. Life—Origin.
I. Sagan, Dorion, 1959- .II. Title.
QH326.M298 1995
577—dc20

Contents

Foreword
Undreamt Philosophies

Why has evolution crafted a sentient species? Why did our consciousness, our realization of our very existence, evolve? What purpose does it serve? I am persuaded by behaviorist Nicholas Humphries' conjecture that, in being able to consult their inner selves, our ancestors gained insight on the minds of their mates, offspring, and other members of their social bands. Knowing thyself is the best way to knowing others, and thus an advantage in negotiating the complexities of daily social life.

We humans are, of course, animals. I have long thought that the very best insight into what it means to be a living, breathing animal is simply to consider one's very own life. However far our cognitive, cultural capacities have taken us from traditional existence within local ecosystems, we nonetheless still obtain energy and food to develop, grow and maintain our corporeal existence. Many of us (perhaps too many of us) also engage in reproduction. As Lynn Margulis and Dorion Sagan tell us in *What Is Life?*, the business of maintaining corporeal existence and reproducing are quintessential activities, the very hallmarks of life. To know oneself as an organism, then, is to establish quite a few of the very basics of all living systems.

But humans, of course, do not constitute the entire biological universe. We are but one species of tens of millions now inhabiting planet Earth. And so we cannot expect to divine all of life's mysteries, all the different nuances of what it means to be alive, simply by consulting our inner selves. There are inherent limits to the revelatory principle of knowing thyself in order to know the world. But even I, a seasoned practitioner in evolutionary biology, was not fully prepared for the wild spectrum of life presented to us by Lynn Margulis and Dorion Sagan in *What Is Life?*. For in these pages we meet organisms vastly different from ourselves. And we encounter ways of thinking about life that could not possibly arise from simple introspection.

What is Life? is a feast of biological and intellectual diversity. Here we meet microbes—microscopic organisms—for which oxygen is a poison, and others who "breathe" sulfur compounds. And still others which feed on hydrogen and carbon dioxide using neither energy from the sunlight, nor that from the flesh of others. We encounter bacteria routinely exchanging genetic materials with other species—even after billions of years of evolutionary separation. We see the entire outer rind of Earth portrayed in convincing fashion as a single, mega-living system. And we learn that the evolutionary process that has produced this prodigious array has done so in astonishing ways—melding separate, simple organisms more than once to produce more complex descendant species. And therein lies a particularly interesting saga of intellectual sleuthing and derring do.

Darwin taught us that all of life is descended from a single common ancestor. In *What Is Life?*, Margulis and Sagan tell us the amazing fact that not only are our own mammalian, nucleated ("eukaryotic") cells descended from ancient bacteria, **they are literally amalgams of several different strains of bacteria.** Amazing! Stranger than fiction! And undreamt of in traditional biological philosophies—until Lynn Margulis began her research a quarter of a century ago.

Lynn Margulis has achieved what every scientist dreams of, but few are destined to accomplish: she has rewritten the basic textbooks. She conceived of a logical, yet audacious explanation of an outstanding fact. Human cells, like those of all animals, the eucalyptus tree and the mushroom, have most, but not all, of their DNA corralled into a cellular nucleus, neatly walled off from the various organelles that dot the plains of their typical cell's cytoplasm. It was the "not all" that attracted her attention: some of these extra-nuclear organelles—specifically, the power plants of all animal and plant cells, the "mitochondria"—were also known to have their own DNA. In plants, both mitochondria and chloroplasts, the locus of photosynthesis, have their own DNA complements. The simple question she faced was: why? Why is there an independent set of genes in these cytoplasmic organelles, when all of the "normal" genetic material is otherwise organized as double sets of chromosomes within the bounds of the nuclear walls?

Biological structures are signals of ancient evolutionary events. We owe the five fingers on our hands not to novel evolutionary events a million years ago on the African savannahs, but rather as a holdover from the original complement of five digits on the forefoot of the earliest land vertebrates ("tetrapods"), who evolved some 370 million years ago.

So, too, is mitochondrial DNA a holdover, a signal of an evolutionary event. But this was like no other event ever proposed in evolutionary annals: Lynn Margulis, to her everlasting credit, saw that separate DNA complements **imply the fusion of at least two different kinds of other organisms, each with its own DNA complement, to form a single, complex "eukaryotic" cell.** Initially condemned as heresy, this elegant idea had so much going for it that the biological world has long since accepted it. There is simply no other plausible explanation for the existence of separate DNA complements in a "single" cell.

In *What Is Life?*, Lynn Margulis and Dorion Sagan tell us precisely which kinds of bacteria fused to form the original nucleated cells—*our* cells. But that is far from all, for the Margulis mind, ever restless, has kept on pushing the envelope. *What Is Life?* presents the case for an even **earlier** evolutionary fusion of bacteria species. Margulis has come to be convinced that such symbiotic origins of novel life forms ("**symbiogenesis**") has been far more common than ever dreamt by evolutionary biologists steeped in the Darwinian tradition—a tradition that emphasizes competition far more than cooperation in the evolutionary process. Symbiogenesis is

Margulis' central contribution to the evolutionary dialogue, which has become enriched through her efforts to see the grand implications latent in the history of the microbial world.

But there is more to the Margulis-Sagan canon than even these profoundly new, and heretofore undreamt, philosophies. Tireless champions of the microbial world, the authors have labored mightily in an almost public-relations sense, striving to reveal the immensely diverse array of micro-organisms. For microbes will not only inherit the earth (should, for example, we complex multicellular creatures fall prey to the next spasm of mass extinction); for microbes got here long before we did, and in a very real sense they already "**own**," and most certainly **run**, the global system. They fix and recycle nitrogen and carbon and other essential elements otherwise unavailable to our bodies; they produce oxygen, natural gas (methane), and so on and on. Without the microbial world, life as we ourselves experience it simply could not be.

All of which lifts the Margulis gaze from the microscopic to the global: Earth truly is a living system, a globally pulsing amalgam of organisms and the physical "inanimate" world. Whether or not one chooses to call this system "Gaia" and pronounce it as alive as any organism does not, in a profound sense, really matter. For in reading **What Is Life?**, we see, clearly and simply, that the global system linking life with the physical realm truly does exist, and that we humans, despite appearances and protestations to the contrary, are still very much a part of that system.

Which takes us back to the ultimate value of being aware of our own existence. As we read **What Is Life?**, we think about life's riotous diversity and evolution's exuberance, and we realize that the global system, all that life, and, in the end, *our* very own existence, are very much under threat—from our very own selves. **What Is Life?** combines the stranger-than-fiction realities of the living world

with the kind of intellectual force that can reveal new undreamt philosophies. It yields the understanding we so desperately need if we are to confront the mounting threat we humans pose to the global ecosystem as we cross over the Millennial divide. Knowledge is power, and **What Is Life?** equips us with an understanding of the living world that we so desperately need if we—along with the world's ecosystems—are to survive.

Niles Eldredge

*American Museum of
Natural History*

Life: The Eternal Enigma

Life is something edible, lovable, or lethal.

—**James E. Lovelock,** inventor

Life is not a thing or a fluid any more than heat is. What we observe are some unusual sets of objects separated from the rest of the world by certain peculiar properties such as growth, reproduction, and special ways of handling energy. These objects we elect to call "living things."

—**Robert Morison,** physicist

In the Spirit of Schrödinger

Half a century ago, before the discovery of DNA, the Austrian physicist and philosopher Erwin Schrödinger inspired a generation of scientists by rephrasing for them the timeless philosophical question: *What Is Life?* [PLATE 3] In his classic 1944 book, bearing that title, Schrödinger argued that, despite our "obvious inability" to define it, life would eventually be accounted for by physics and chemistry. Life, Schrödinger held, is matter which, like a crystal—a strange, "aperiodic crystal"— repeats its structure as it grows. But life is far more fascinating and unpredictable than any crystallizing mineral:

> The difference in structure is of the same kind as that between an ordinary wallpaper in which the same pattern is repeated again and again in regular periodicity and a masterpiece of embroidery, say a Raphael tapestry, which shows no dull repetition, but an elaborate, coherent, meaningful design traced by the great master.[1]

Schrödinger, a Nobel laureate, revered life in all its marvelous complexity. Indeed, although he devised the wave equation that helped give quantum mechanics theory a firm mathematical basis, he never conceived of life as simply a mechanical phenomenon.

Our book, addressing life's fulness without sacrificing any science, reproduces not only Schrödinger's title but also, we hope, his spirit. We have tried to put the life back into biology.

What is life? is surely one of the oldest questions. We live. We—people, birds, flowering plants, even algae glowing in the ocean at night—differ from steel, rocks, inanimate matter.

We are alive. But what does it mean to live, to be alive, to be a discrete being at once part of the universe but separated from it by our skin? What is life?

Erwin Shrödinger: a physicist whose emphasis on the physiochemical nature of life helped inspire the discovery of DNA and the molecular biological revolution.

Thomas Mann gave an admirable, if literary, answer in the novel *The Magic Mountain*:

> What was life? No one knew. It was undoubtedly aware of itself, so soon as it was life; but it did not know what it was ... it was not matter and it was not spirit, but something between the two, a phenomenon conveyed by matter, like the rainbow on the waterfall, and like the flame. Yet why not material—it was sentient to the point of desire and disgust, the shamelessness of matter become sensible of itself, the incontinent form of being. It was a secret and ardent stirring in the frozen chastity of the universal; it was a stolen and voluptuous impurity of sucking and secreting; an exhalation of carbonic gas and material impurities of mysterious origin and composition.[2]

Our ancestors found spirits and gods everywhere, animating all of nature. Not only were the trees alive but so was the wind howling across the savanna. Plato, in his dialogue *Laws*, said that those perfect beings, the planets, travel around the earth voluntarily in circles. Medieval Europeans believed the microcosm, the small world of the person, mirrored the macrocosm, the universe; both were part matter and part spirit. This ancient view lingers in the animals of the zodiac and in the astrological notion that celestial bodies influence mundane ones.

In the seventeenth century the German astrologer-astronomer Johannes Kepler (1571-1630) calculated that planets travel around the sun in ellipses. Nevertheless, Kepler (who wrote the first work of science fiction and whose mother was arrested as a witch) believed that the stars inhabit a three-kilometer-thick shell far beyond the solar system. He considered Earth a breathing, remembering, habit-forming monster. Although Kepler's view of a living Earth now seems whimsical, he reminds us that science is asymptotic: it never arrives at but only approaches the tantalizing goal of final knowledge. Astrology gives way to astronomy; alchemy evolves into chemistry. The science of one age becomes the mythology of the next. How will future thinkers assess our own ideas? This movement of thought—of living beings questioning themselves and their surroundings—is at the heart of the ancient question of what it means to be alive.

Life—from bacterium to biosphere—maintains by making more of itself. We focus on self-maintenance in our first chapter. Next, we trace views of life from very early on through European mind-body dualism and then to modern scientific materialism. Chapter 3 explores life's origins and its memory-like preservation of the past. Our ancestors—the bacteria that brought Earth's surface to life—are featured in chapter 4.

Through symbiotic mergers, bacteria evolved into the protists of chapter 5. Protists are unicells, including algae, amebas, ciliates, and other postbacterial cells with erotic habits anticipating our own; they evolved into multicelled beings experiencing sex and death. We call the unicellular protists, together with their close multicellular relatives—some of which are very large—protoctists. The bacteria that formed protoctists were to have a spectacular future. They became animals (chapter 6), fungi (chapter 7), and plants (chapter 8). In the last chapter we pursue the unorthodox but common-sensical idea that life—not just human life but all life—is free to act and has played an unexpectedly large part in its own evolution.

Life's Body

Life, although material, is inextricable from the behavior of the living. Defying definition—a word that means "to fix or mark the limits of"—living cells move and expand incessantly. They overgrow their boundaries; one becomes two become many. Although exchanging a great variety of materials and communicating a huge quantity of information, all living beings ultimately share a common past.

Perhaps even more than Schrödinger's "aperiodic crystal," life resembles a fractal—a design repeated at larger or smaller scales. [PLATE 4] Fractals, beautiful for their delicacy and surprising in their apparent complexity, are produced by computers, as graphics programs iterate, or repeat, a single mathematical operation thousands of times. The "fractals" of life are cells, arrangements of cells, many-celled organisms, communities of organisms, and ecosystems of communities. Repeated millions of times over thousands of millions of years, the processes of life have led to the wonderful, three-dimensional patterns seen in organisms, hives, cities, and planetary life as a whole.

Life's body is a veneer of growing and self-interacting matter encasing Earth. Twenty kilometers thick, its top is the atmosphere and its bottom is continental rock and ocean depths. Life's body is like a tree trunk. Only its outermost tissues grow. Unless protected by technology, itself an extension of life, any individual removed from the living sphere is doomed.

Life, as far as is known, is limited to the surface of this third planet from the sun. Moreover, living matter utterly depends on this sun, a medium-sized star in the outback of the Milky Way Galaxy. Less than one percent of the solar energy that strikes the earth is diverted to living processes. But what life does with that one percent is astounding. Fabricating genes and offspring from water, solar energy, and air, festive yet dangerous forms mingle and diverge, transform and pollute, slaughter and nurture, threaten and overcome. Meanwhile, the biosphere itself, subtly changing with the comings and goings of individual species, lives on as it has for more than three thousand million years.

Animism vs. Mechanism

If you wish, you can reach for a glass of water or snap this book shut. From the experience of willing our bodies to move came animism: the view that winds come and go, rivers flow, and celestial bodies guard the heavens because something inside each wills the movement. In animism all things, not just animals, are seen to be inhabited by an inner, animating spirit. Formalized in polytheistic religion, the multiplicity of gods—a moon god, earth god, sun god, wind god, and so on—was replaced in Islam, Judaism, and Christianity by a single god who crafted the world. Winds and rivers and celestial bodies lost their will, but living organisms—especially humans—retained theirs.

Finally, the last outposts of animism—living organisms—yielded to the philosophy of mechanism. Motion need not imply any inner consciousness; the program could have been "built in" by a creator. Wind-up toys and automated models of the solar system suggested to their inventors that even living things may be constructible from lifeless mechanisms, subtle concealed springs, tiny unseen pulleys, levers, cogs, and gears. Comparing flowing

4

Like Julia and Mandelbrot patterns, Peter Allport's duality fractal exemplifies a crucial trait of life: the iterative production of complex entities from a component whose design repeats at ever greater scales. Whereas computer graphics are due to the repetition of an algorithm, the motive force behind "living fractals" is the reproduction of cells.

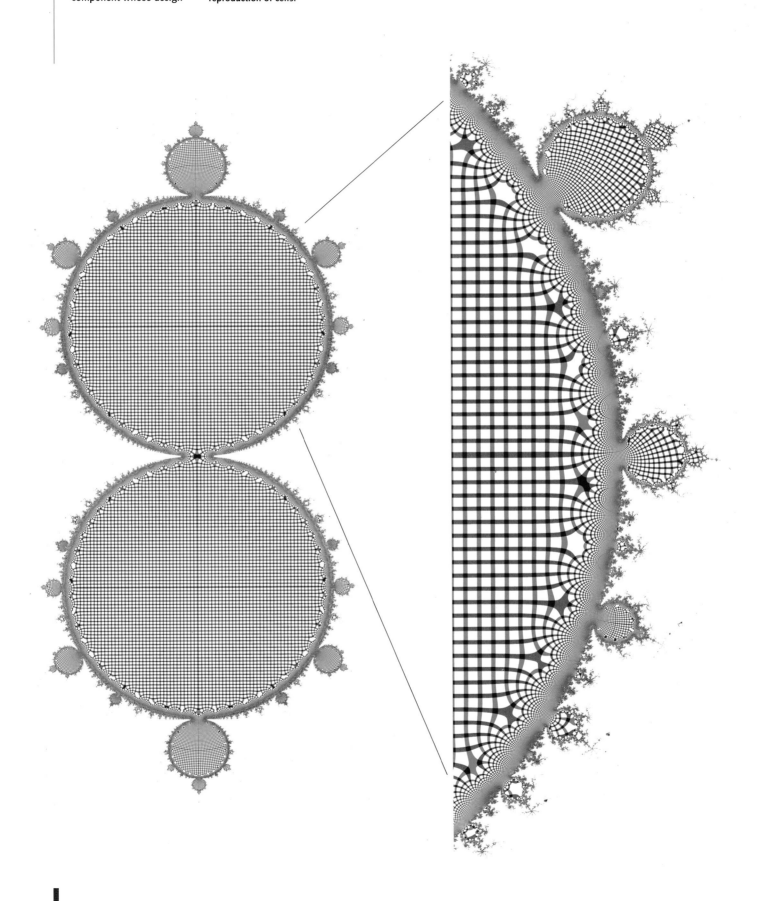

blood to a hydraulic system, the heart to a pump, English physician William Harvey (1578-1647) discovered circulation of the blood. Scientists sleuthed out the world's secret mechanisms, part of an overall design. Natural History revealed the world to be a giant mechanism made according to the mind of God.

Isaac Newton (1642-1727) became the high priest of mechanism. A devoted student of alchemy, scripture, and the occult, Newton made unparalleled innovations in optics, physics, and mathematics. In doing so he helped bridge the gap from the medieval cosmos to the modern one. Explaining the motions of the planets with a new law of gravity, Newton's equations showed that the world of the heavens and that of Earth were one and the same; the force that kept the moon in orbit was also the force that thuds an apple to the ground. So revealing were Newton's discoveries of "laws" governing the entire universe that to some it seemed he had—in Kepler's words— "glimpsed the mind of God." Inspired by Newton's analyses, (Pierre-Simon) marquis de Laplace (1749-1827) speculated that, with sufficient information, the entire future of the universe, even the most minute human action, could be predicted. Far from being moved by hidden spirits, the celestial bodies now seemed to be under the governance of preexistent mathematical laws. Divine intervention became increasingly superfluous. God did not need to fiddle with creation. He had crafted it to last. The cosmos worked itself.

With a grasp of gravitation's cosmic sweep, scientists were spurred to explore phenomena once considered beyond the ken of human comprehension. Electricity and magnetism, chemicals and colors, radiation and heat, explosion and chemical change were all described with an eye to their underlying unity. Optical instruments, telescope and microscope, presented formerly unseen worlds of the very far and the very near. Experiment and criticism replaced blind acceptance of classical authority and divinely revealed truth. Scientists coaxed nature to yield some of her most private secrets. Oxygen's role in fire, lightning as electrical discharge, gravity as the invisible force causing the tides and attracting the moon into Earth's orbit— one by one nature laid down her cards.

Under the spell of the mechanical world view, the ancient alchemical dream of shaping nature to human will became technological reality. After centuries of meddling with steamy concoctions in a Faustian quest to be godlike, a 1953 discovery seemed to reveal the very secret of life. Life was chemical and the material basis of heredity was DNA, whose helical and staircase-like structure made clear how molecules copied themselves. Indeed, the "aperiodic crystal" that Schrödinger had predicted was uncannily similar to the double helix first described by the English chemist Francis Crick and American whiz kid James D. Watson. Replication was no longer beholden to a mysterious "vital principle"; it was the straightforward result of interacting molecules. The description of how DNA fabricated a copy of itself out of ordinary carbon, nitrogen, and phosphorus atoms was perhaps the most spectacular of all mechanism's successes. But paradoxically, this success born of self-directed minds seemed to portray life—including the scientists themselves—as the result of atoms involuntarily interacting according to changeless and inviolable chemical law.

Between these two extremes—the entire universe as alive, and the living organism as chemical and physical machine—lies the panorama of opinion. But is there not something wrong with both the mechanization of life *and* the vitalization of matter?

The world as a vast machine fails to account for our own self-awareness and self-determination because the mechanical worldview denies choice. Mechanisms, after all, don't act; they react. And mechanisms, moreover, don't come into existence on their own. The assumption that the universe is a mechanism implies that it was made according to some humanlike design—that is, by some living creator. In other words, successful as it is, the scientific mechanistic worldview is deeply metaphysical; it is rooted in religious assumptions.

The animistic view of the cosmos as a huge organism is also flawed. It blurs the distinctions among what is living, what is dead, and what has never been alive. If everything were alive, there would be no interest in—and scientists never would have discovered the replicative chemistry of—life.

We thus reject mechanism as naive and animism as unscientific. Even so, life, as an emergent behavior of matter and energy, is best known by science. Schrödinger was correct in advocating a search for the physico-chemical underpinnings of life. So are Watson and Crick and other physicists and molecular biologists who hail the structure of DNA as a key to life's secrets. Like an uncoiling spring pushing the soft gears of life, DNA copies itself as it directs the making of proteins that together form the leopard's spots, the spruce tree's cone, and living bodies in general. Understanding how DNA works may be the greatest scientific breakthrough in history. Nonetheless, neither DNA nor any other kind of molecule can, by itself, explain life.

Janus Among the Centaurs

The American architect Buckminster Fuller applied "synergy" (from Greek *synergos*, working together) to describe entities that behave as more than the sum of their parts. From a scientific standpoint, life, love, and behavior appear to be synergistic phenomena. When certain chemicals—in water and in oil—came together long ago, life was the result. Synergy also fits the emergence of protist cells from bacteria, and of animals from such cells.

The common view is that life evolves by random genetic change that is, moreover, detrimental more often than not. Chance mutations, blind and undirected, are touted to lead to evolutionary novelty. We (and a growing contingent of like-minded students of life) do not entirely agree. Great gaps in evolution have been leaped by symbiotic incorporation of previously refined components—components that have been honed in separate lineages. Evolution doesn't start anew each time a new life form appears. Preexisting modules (which turn out to be primarily bacteria), already generated by mutation and retained by natural selection, come together and interface. They form alliances, mergers, new organisms, whole new complexes that act and are acted on by natural selection.

But natural selection by itself cannot generate any evolutionary innovation, as Charles Darwin was well aware. Natural selection, rather, relentlessly preserves the former refinements and newly generated novelty by culling those less able to live or reproduce. Biotic potential—life's tendency to reproduce as much as possible—takes care of the rest. But first, novelty must arise from somewhere. In synergy two distinct forms come together to make a surprising new third one.

Cowboys, for example, settled the American West. Some native Americans perceived the human-horse invaders as centaurs—two-headed, multilimbed beings. The novelist and philosopher Arthur Koestler (1905-1983) has called the coexistence of smaller beings in larger wholes "holarchy."[3] Most people, by contrast, think that life on Earth is hierarchical, a great chain of being with humans on top. Koestler's coinage is free of implications of "higher" or that one of the constituents in the holarchy is somehow controlling the others. The constituents, too, were given a new name by Koestler. Not merely parts, they are "holons"—wholes that also function as parts.

In his metaphysical as well as terminological rethinking, Koestler invoked the double-faced Janus, who in Roman mythology was the god of portals and the patron of beginnings and endings. In our view, just as Janus simultaneously looks backward and forward, so humans are not at the height of creation but point dually to the smaller realm of cells and the larger domain of biosphere. Life on Earth is not a created hierarchy but an emergent holarchy arisen from the self-induced synergy of combination, interfacing, and recombination.

Blue Jewel

The best part of a journey can be returning. By sending monkeys and cats into orbit, people to the moon, and robots to Venus and Mars, humankind has developed a new respect for, and a deeper understanding of life on Earth.

In 1961 the Soviet Union's *Vostok I* carried the first human into orbit around Earth. Since then, gazing "down" at this turquoise orb — venturing out on a spacewalk as if about to jump from the world's highest diving board — cosmonauts and astronauts have groped for words that do justice to their experience. Eugene A. Cernan, an astronaut of both the Gemini and Apollo lunar missions, and the last person to walk on the moon, describes the view:

> When you are in Earth orbit looking down you see lakes, rivers, peninsulas....You quickly fly over changes in topography, like the snow-covered mountains or deserts or tropical belts — all very visible. You pass through a sunrise and sunset every ninety minutes. When you leave Earth orbit...you can see from pole to pole and ocean to ocean without even turning your head....You literally see North and South America go around the corner as Earth turns on an axis you can't see and then miraculously Australia, then Asia, then all of America comes to replace them.... You begin to see how little we understand of time....You ask yourself, where am I in space and time? You watch the sun set over America and rise again over Australia. You look back "home"...and don't see the barriers of color, religion, and politics that divide up this world.[4]

Imagine yourself in orbit. Circling the planet every ninety minutes, the earthly experience of time and space undergo a mutual metamorphosis. Gravity lessens; up and down become relative. Day follows night in a patchwork blend. The sun cuts through the thin ribbon that is the atmosphere, flooding the cabin of the spacecraft from red to green to purple, through all the colors of the rainbow. You are plunged into black. Earth becomes the place where there are no stars. If Earth can be seen at all it is as a flicker of tiny lights — cities — on the surface of the sun-eclipsing globe. "Day" breaks again, revealing the cloud-flecked blue ocean. Jettisoned into a hyperperspective, the sky is now below. As if floating dreamily away from your own body, you watch the planet to which you are now tied by only the invisible umbilicus of gravity and telecommunication.

The act of viewing Earth from space echoes that of a baby glimpsing, and really seeing, itself in a mirror for the first time. [PLATE 5] The astronaut gazes upon the body of life as a whole. The French pyschoanalyst Jacques Lacan posits a stage in human development called "the mirror stage."[5] The infant, unable to control its limbs, looks into the mirror and perceives its whole body. Humanity's jubilant perception of the global environment represents the mirror stage of our entire species. For the first time we have caught a glimpse of our full, planetary form. We are coming to realize that we are part of a global holarchy that transcends our individual skins and even humanity as a whole.

Television images in 1969 revealed astronauts bounding over the lunar dust. The moon, once a synonym for the unattainable, was reached. A cratered wasteland, bone-dry, the moon was nevertheless still daunting in its lifelessness. As the cosmic perspective was broadcast, we homebodies were given a futuristic ride and were offered a new view of the world, a new worldview with the power to rally Earth's peoples around an icon more potent than any flag. Members of disparate religious and

A composite satellite photo showing visible evidence of night life on Earth: city lights. In orbit it becomes clear that night is shadow.

spiritual traditions could now join together as citizens of Earth. Individuals so affected, those who saw the potential, came to know that the whole former understanding of life was parochial, a result of where we lived. Even time was upset: night became shadow.

Tribal conflicts, national politics, and the colored geographic regions of maps are invisible from space. Science has, of course, revealed to us that this blue jewel orbits but a lackluster star in the outskirts of a spiral galaxy with myriad stars within a universe of myriad galaxies. All our history and civilization has transpired under the gaseous blanket of, really, a middling planet in one solar system. Voyaging in space, we saw Earth is home. But it is more than home: it is part of us. In contrast with the pale moon in the dead solar system of our galactic suburbs, this third planet from the sun, our Earth, is a blue-and-white flecked orb that looks alive.

Is There Life on Mars?

Unexpectedly, the search for life on Mars provided scientific confirmation of the "body" of life as a whole on Earth. The Viking mission, launched in 1975, sent two orbiters and two landers to Mars.

Although returning spectacular images of "Marscapes," the Viking landers performed a series of experiments that failed to find any evidence of Martian life. Channels carved by ancient rivers were seen, fueling hopes that evidence for past life may yet be found on the red planet.

One scientist, however, was able to search for life on Mars before the Viking mission was launched. In 1967 James E. Lovelock, English inventor of a device that measures chlorofluorocarbons implicated in the production of ozone holes, was consulted by the National Aeronautics and Space Administration (NASA) in its search for extraterrestrial life. NASA was interested in what Lovelock's invention, a gas-measuring instrument some thousand times more sensitive to certain atmospheric constituents than any previous device, might reveal about Mars. An atmospheric chemist, Lovelock suspected that, in principle, life on any planet could be detected by the chemical markers left in that planet's air. Because the constituents of Mars's atmosphere were already known by the spectroscopic signature of the planet's reflected light, Lovelock believed the data already sufficient to determine whether Mars was a living planet. His conclusion: Mars was devoid of life.

Indeed, he boasted with his own brand of quiet iconoclastic mischief that his prediction precluded any need to visit Mars at all and that he could save NASA a prodigious sum of money.

Lovelock had measured Earth's atmospheric gases with a chromatograph outfitted with his new supersensitive "electron capture device." He was startled: the chemistry of Earth's atmosphere, not at all like the atmospheres of Mars and Venus, is utterly improbable. He found that methane, a chief constituent of natural gas and present in the atmospheres of the four giant planets (Jupiter, Saturn, Uranus, and Neptune), freely coexisted in Earth's atmosphere with oxygen at concentrations 10^{35} times higher than expected.

Methane exists at only one to two parts per million in Earth's atmosphere, but even that minuscule proportion is far too high. Methane (one carbon atom surrounded by four hydrogen atoms) and oxygen gas (two oxygen atoms) react explosively with each other to generate heat, producing carbon dioxide and water. Oxygen, the second most abundant gas in the atmosphere, should thus react immediately with methane to make the latter undetectable. Perhaps in the next minute you will die of asphyxiation because all the oxygen atoms will gather in one corner of the room and your brain will be deprived of its absolute requirement for oxygen gas. Such a calamity is improbable to the point of absurdity. Yet the chemical mixture of methane and oxygen in the Earth's air is equally freakish. Indeed, not only methane but many other gases in our air should not be detectable, given standard rules of chemical mixing. Given their tendency to react with oxygen, some of our atmosphere's components—methane, ammonia, sulfur gases, methyl chloride, and methyl iodide—are far from chemical equilibrium. Carbon monoxide, nitrogen, and nitrous oxide are respectively ten, ten thousand million, and ten trillion times more abundant than chemistry alone can account for.

Biology, however, offers an answer. Lovelock realized that methane-producing bacteria release this gas in globally significant amounts. Cows, for instance, contribute methane by belching. Belched methane does react with oxygen but, before it disappears, more is produced. The methane is made from grass by protists and bacteria in the cow's rumen, a special stomach.

Life has made our atmosphere chemically reactive and orderly, while exporting heat and disorder to space. Lovelock maintained that the atmosphere is as highly ordered as a painted tortoise's shell or a sand castle on a deserted beach. And life's inveterate ordering has left its traces on other planets. On 20 July 1976 a lander was left on Mars by the 3.6 metric ton *Viking I* spacecraft. [PLATE 6] Although not what scientists were looking for, this machine, sitting 571 million kilometers away at Chryse Planitia on red sand, is the best, and so far the only evidence of life on Mars: solar system exploring, technological human life.

Life as Verb

Lovelock's analyses have pushed biologists to realize that life is not confined to the things now called organisms. Self-transforming, holarchic life "breaks out" into new forms that incorporate formerly self-sufficient individuals as integral parts of greater identities. The largest of these levels is the planetary layer, the biosphere itself. Each level reveals a different kind of "organic being." This is the term that Darwin used throughout his opus, *The Origin of Species*. ("Organism," like "scientist" and "biology," had not yet been coined.) "Organic being" merits resurrection as it affords the recognition that a "cell" and the "biosphere" are no less alive than an "organism."

Life—both locally, as animal, plant, and microbe bodies, and globally, as the biosphere—is a most intricate material phenomenon. Life shows the usual chemical and physical properties of matter, but with a twist. Beach sand is usually silicon dioxide. So are

Mar's Utopian Plain as
seen by Viking II which
landed September 3, 1976.
The blue-star field and red
stripes of the USA's flag
help corroborate the
chromatic correctness of
the salmon-colored sky
caused by dust particles
suspended in the Martian
atmosphere.

the innards of a mainframe computer—but a computer isn't a pile of sand. Life is distinguished not by its chemical constituents but by the behavior of its chemicals. The question "What is life?" is thus a linguistic trap. To answer according to the rules of grammar, we must supply a noun, a thing. But life on Earth is more like a verb. It repairs, maintains, recreates, and outdoes itself.

This surge of activity, which not only applies to cells and animals but to Earth's entire atmosphere, is intimately connected to two of science's most famous laws—the laws of thermodynamics. The first law says that throughout any transformation the total energy of any system and its environment is neither lost nor gained. Energy—whether as light, movement, radiation, heat, radioactivity, chemical or other—is conserved.

But not all forms of energy are equal; not all have the same effect. Heat is the kind of energy to which other forms tend to convert, and heat tends to disorganize matter. The second law of thermodynamics says that physical systems tend to lose heat to their surroundings.

The second law was conceived during the Industrial Revolution, when the steam engine represented the state-of-the-art in engineering. French physicist Nicolas Carnot (1796–1832), aiming to improve the efficiency of the steam engine (whose governor mechanism was invented by James Watt), came to realize that heat was associated with the movement of minute particles. And from that, he envisioned the principle that is now known as the second law: In any moving or energy-using system entropy increases.

In systems undergoing change, such as steam engines or electric motors, a certain amount of the total energy available is already in (and more is converted into) a form that is unavailable for useful work. Although the amount of energy in the system and its environment stays the same (i.e., the first law of thermodynamics, of conservation of energy, holds), the amount of energy available to do work decreases. In computer science entropy is measured as the uncertainty in the information content of a message. The second law unequivocally claims that in changing systems entropy increases, implying that heat, noise, uncertainty, and other such forms of energy not useful for work, increase. As local systems lose heat, the universe as a whole is gaining it. Although not so popular now, in the past physicists and chemists have made the prediction that the universe will whimper out in a "heat death" as a consequence of the tendency for entropy to increase. More recently, they have even invented the word "negentropy" for life, which, in its tendency to increase information and certainty, seems to contradict the second law. It doesn't; the second law holds as long as one regards the system (life) in its environment.

In steam engines, coal was burned and carbon joined with oxygen, a reaction that, generating heat, made machine parts move. The left-over heat that was generated was unusable. The heat in a cabin on a snow-covered mountain seeks with seeming purpose any available crack or opening to mix with the cold air outside. Heat naturally dissipates. This dissipative behavior of heat illustrates the second law: the universe tends toward an increase in entropy, toward even temperatures everywhere, as all the energy transforms into useless heat spread so evenly that it can do no work. Heat dissipation, we are usually told, results from random particle motion. But there are other interpretations.

Some scientists have begun to interpret the second law's predilection for heat-energy as the basis for apparent purposeful action. Ilya Prigogine, a Belgian Nobel laureate, helped pioneer the consideration of life within a larger class of "dissipative structures," which also includes decidedly nonliving centers of activity like whirlpools, tornadoes, and flames.[6] A rather awkward term because it focuses on what the structures—actually, systems, not structures—throw away rather than what they retain and build, a dissipative system maintains itself (and may even grow) by importing "useful" forms of energy and exporting, or dissipating, less useful forms—notably, heat. This thermodynamic view of life actually goes back to Schrödinger, who also likened living beings to flames, "streams of order" that maintain their forms.

American scientist Rod Swenson has argued that the seeming purpose displayed in heat's tendency to dissipate with time is intimately related to the behavior of life forms striving to perpetuate themselves. In Swenson's view, this entropic universe is pocked by local regions of intense ordering (even life) because it is through ordered, dissipative systems that the rate of entropy production in the universe is maximized. The more life in the universe, the faster that various forms of energy are degraded into heat.[7]

Swenson's view shows how life's seeming purpose—its seeking behavior, its directedness, which philosophers call teleology—is related to the behavior of heat. Scientists do not as a rule endorse teleology. They consider it unscientific, a holdover from the primitive days of animism. Teleology is nevertheless embedded in language, and it cannot and need not be eliminated from the sciences. The prepositions "to" and "for," which build teleology—that is, purposefulness—into language, speak of a future-directedness that seems present, to some degree, in all living beings. One should not assume that only humans are future-oriented. Our own frenetic attempts (and those of the rest of life) to survive and prosper may be a special, four-billion-year old way the universe has organized itself "to" obey the second law of thermodynamics.

Self-Maintenance

Islands of order in an ocean of chaos, organisms are far superior to human-built machines. Unlike James Watts's steam engine, for example, the body concentrates order. It continuously self-repairs. Every five days you get a new stomach lining. You get a new liver every two months. Your skin replaces itself every six weeks. Every year, ninety-eight percent of the atoms of your body are replaced. This nonstop chemical replacement, metabolism, is a sure sign of life. This "machine" demands continual input of chemical energy and materials (food).

Chilean biologists Humberto Maturana and Francisco Varela see in metabolism the essence of something quite fundamental to life. They call it "autopoiesis." Coming from Greek roots meaning self (*auto*) and making (*poiein*, as in "poetry"), autopoiesis refers to life's continuous production of itself.[8] Without autopoietic behavior, organic beings do not self-maintain—they are not alive.

An autopoietic entity metabolizes continuously; it perpetuates itself through chemical activity, the movement of molecules. Autopoiesis entails energy expenditure and the making of messes. Autopoiesis, indeed, is detectable by that incessant life chemistry and energy flow which is metabolism. Only cells, organisms made of cells, and biospheres made of organisms are autopoietic and can metabolize.

DNA is an unquestionably important molecule for life on Earth, but the molecule itself is not alive. DNA molecules replicate but they don't metabolize and they are not autopoietic. Replication is not nearly as fundamental a characteristic of life as is autopoiesis. Consider: the mule, offspring of a donkey and a horse, cannot "replicate." It is sterile, but it metabolizes with as much vigor as either of its parents; autopoietic, it is alive. Closer to home, humans who no longer, never could, or simply choose not to reproduce can not be relegated, by the strained tidiness of biological definition, to the realm of the nonliving. They too are alive.

In our view, viruses are not. They are not autopoietic. Too small to self-maintain, they do not metabolize. Viruses do nothing until they enter an autopoietic entity: a bacterial cell, the cell of an animal, or of another live organism. Biological viruses reproduce within their hosts in the same way that digital viruses reproduce within computers. Without an autopoietic organic being, a biological virus is a mere mixture of chemicals; without a computer, a digital virus is a mere program.

Smaller than cells, viruses lack sufficient genes and proteins to maintain themselves. The smallest cells, those of the tiniest bacteria (about one ten-millionth of a meter in diameter) are the minimal autopoietic units known today. Like language, naked DNA molecules, or computer programs, viruses mutate and evolve; but, by themselves, they are at best chemical zombies. The cell is the smallest unit of life.

When a DNA molecule produces another DNA molecule exactly like itself, we speak of replication. When living matter, as a cell or as a body made of cells, grows another similar being (with differences attributable to mutation, genetic recombination, symbiotic acquisition, developmental variation, or other factors), we speak of reproduction. [PLATE 7] When living matter continues to reproduce altered forms that, in turn, make altered offspring, we speak of evolution: change in populations of life forms over time. As Darwin and his legacy stress, more reproducing cells and bodies are produced by budding, cell division, hatching, birth, spore forma-tion, and the like than can ever survive. Those that cope long enough to reproduce are "naturally selected." More bluntly, survivors are not so much selected for their success as those failing to repro-duce before dying are selected against.

Identity and self-maintenance require metabolism. Metabolic chemistry (often called physiology) precedes reproduction and evolution. For a population to evolve, its members must reproduce. Yet before any organic being can reproduce, it must first self-maintain. Within the lifetime of a cell, each of five thousand or so different proteins will completely interchange with

7
Pachnoda. **Phylum: Arthropoda. Kingdom: Animalia. Close-up of the intestine of the beetle larva showing a tree-like organ that houses methanogenic bacteria. Living in the intestine for thousands of generations, the methangens have not only found a home but produced this symbiotic "beetle" organ.**

the surroundings thousands of times. Bacterial cells produce DNA and RNA (nucleic acids), enzyme proteins, fats, carbohydrates, and other complex carbon chemicals. Protist, fungi, animal, and plant bodies all produce these and other substances as well. But most importantly, and amazingly, any living body produces itself.

This energetic maintenance of unity while components are continuously or intermittently rearranged, destroyed and rebuilt, broken and repaired is metabolism, and it requires energy. In accordance with the second law, autopoietic self-maintenance preserves or increases internal order only by adding to the "disorder" of the external world, as wastes are excreted and heat is vented. All living beings must metabolize and therefore all must create local disorder: useless heat, noise, and uncertainty. This is autopoietic behavior, reflecting the autopoietic imperative required for any organic being that lives, that continues to function.

The autopoietic view of life differs from standard teachings in biology. Most writers of biology texts imply that an organism exists apart from its environment, and that the environment is mostly a static, unliving backdrop. Organic beings and environment, however, interweave. Soil, for example, is not unalive. It is a mixture of broken rock, pollen, fungal filaments, ciliate cysts, bacterial spores, nematodes, and other microscopic animals and their parts. "Nature," Aristotle observed, "proceeds little by little from things lifeless to animal life in such a way that it is impossible to determine the exact line of demarcation."[9] Independence is a political, not a scientific, term.

Since life's origin, all living beings, directly or circuitously, have been connected, as their bodies and populations have grown. Interactions occur, as organisms connect via water and air. Darwin, in his *Origin of Species*, likened the complexity of these interactions to "an entangled bank"—too complex for us humans even to begin to sort out: "Throw

up a handful of feathers, and all fall to the ground according to definite laws; but how simple is the problem where each shall fall compared to that of the action and reaction of the innumerable plants and animals." Yet it is the sum of these uncountable interactions that yields the highest level of life: the blue biosphere, in all the holarchic coherence and mysterious grandeur of its budding in and from the black cosmos.

The Autopoietic Planet

The biosphere as a whole is autopoietic in the sense that it maintains itself. One of its vital "organs," the atmosphere, is clearly tended and nurtured. Earth's atmosphere, approximately one-fifth oxygen, differs radically from those of Mars and Venus. The atmospheres of these planetary neighbors are nine parts in ten carbon dioxide; in Earth's atmosphere, carbon dioxide accounts for only three parts in ten thousand. If Earth's biosphere were not made of carbon dioxide-consuming beings (plants, algae, photosynthetic, and methane-producing bacteria, among myriad other life forms), our atmosphere would long ago have reached a carbon dioxide-rich chemical balance. And virtually every molecule capable of reacting with another molecule would already have reacted. Instead, the combined activities of autopoietic surface life have led to an atmosphere in which oxygen has been maintained at levels of about twenty percent for at least 700 million years. [CHART A]

Other evidence for life on a planetary scale comes from astronomy. According to standard astrophysical models of the evolution of stars, the sun used to be cooler than it is now. The sun's luminosity has increased by thirty percent or more since life began on Earth. Living things can grow and reproduce only in a limited temperature range within which water is liquid. Fossils of life more than three thousand million years old confirm that ancient temperatures were not all that dissimilar

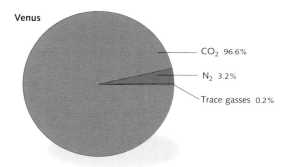

Venus

CO₂ 96.6%

N₂ 3.2%

Trace gasses 0.2%

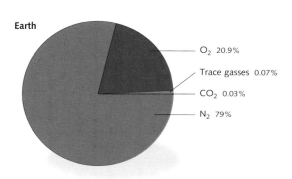

Earth

O₂ 20.9%

Trace gasses 0.07%

CO₂ 0.03%

N₂ 79%

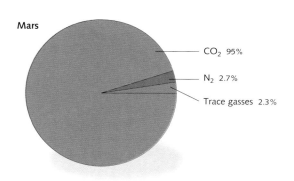

Mars

CO₂ 95%

N₂ 2.7%

Trace gasses 2.3%

Chart A

Atmospheric comparison of Earth and its two planetary neighbors. Note the comparatively high concentration of the explosive gas oxygen and the very low concentration of carbon dioxide on Earth.

This atmospheric anomaly results from the incessant activity of gas-exchanging organisms. The minute physiology of the cell over geological time becomes magnified into the global physiology of the biosphere.

from those prevailing today; other geological evidence suggests that liquid water was widespread on Earth at least four thousand million years ago. The increase in the luminosity of the sun should have dramatically increased the surface temperature of Earth since those early times. Because no dramatic increase has occurred—indeed, the trend may have been a cooling—it appears that the temperature of the entire biosphere has been self-maintained. By responding, life seems to have succeeded in cooling the planetary surface to counter, or more than counter, the overheating sun. Mainly by removing from the atmosphere green-house gases (such as methane and carbon dioxide) which trap heat, but also by changing its surface color and form (by retaining water and growing slime), life responded to prolong its own survival.

Oceanography provides still another glimpse of the body of life as a whole. Chemical calculations suggest that salts should accumulate in the oceans to concentrations perilous to nonbacterial forms of life. Salts, such as sodium chloride and magnesium sulfate, are continuously eroded from the continents and carried into the oceans by rivers. World oceans have, however, remained hospitable to salt-sensitive organisms for at least two thousand million years. Seafaring microorganisms may therefore be sensing and stabilizing ocean acidity and salinity levels on a global scale. How life removes salt from marine waters is obscure. Perhaps salt concentrations too high for most life are lowered, in part, by the vigorous pumping of sodium, calcium, and chloride out of cells and, in part, by formation of evaporite flats. These encrusted fields are rich in sea salt and salt-loving microbes. They often form behind lagoonal barriers made by animals such as corals or when shifting sands are trapped by the mucus and slime formed by microbial communities. Continuous desalination, if it exists, may be part of a global physiology.

Some evolutionary biologists have suggested that

Earth life in its totality cannot constitute a living body, cannot be a living being, because such a body could only have evolved in competition with other bodies of the same sort—presumably, other biospheres. But, in our view, autopoiesis of the planet is the aggregate, emergent property of the many gas-trading, gene-exchanging, growing, and evolving organisms in it. As human body regulation of temperature and blood chemistry emerges from relations among the body's component cells, so planetary regulation evolved from eons of interactions among Earth's living inhabitants.

Using the energy of sunlight, only green plants, algae, and certain green- and purple-colored bacteria can convert compounds from surrounding water and air into the living stuff of their bodies. This sun-energized process, photosynthesis, is the nutritional basis for all the rest of life. Animals, fungi, and most bacteria feed on the purple and green producers. Photosynthesis evolved in microbes soon after the origin of life. At every level, from microbe to planet, organic beings use air and water or other organic beings to build their reproducing selves. Local ecology becomes global ecology. As a corollary, and in spite of English grammar, life does not exist *on* Earth's surface so much as it *is* Earth's surface.

Life extends over the planet as a contiguous, but mobile, cover and takes the shape of the underlying Earth. Life, moreover, enlivens the planet; Earth, in a very real sense, is alive. This is no vague philosophical claim but rather a physiological truth of our lives. Organisms are less self-enclosed, autonomous individuals than communities of bodies exchanging matter, energy, and information with others. Each breath connects us to the rest of the biosphere, which also "breathes," albeit at a slower pace. The biosphere's breath is marked daily by increasing carbon dioxide concentrations on the dark side of the globe and decreasing concentrations on the

lighted side. Annual breathing is marked by the passage of the seasons; photosynthetic activity kicks up in the northern hemisphere just as it is winding down in the Southern.

Taken at its greatest physiological extent, life *is* the planetary surface. Earth is no more a planet-sized chunk of rock inhabited with life than your body is a skeleton infested with cells. [CHART B]

Seasonal Fluctuations of Carbon Dioxide in the Northern Hemisphere
(Measured in parts per million.)

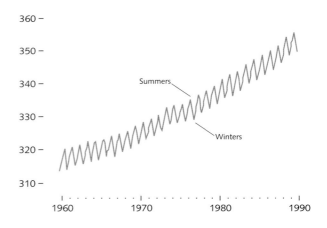

Chart B
The peaks of the zigzags represent increase in atmospheric carbon dioxide during summers in the northern hemisphere; the overall upward trend indicates rising levels of CO_2 due to industrial activity. This seasonal and annual fluctuation of carbon dioxide in the Earth's atmosphere attests to "breathing" on a global scale. The total carbon dioxide increase may, by the greenhouse effect, raise planetary temperatures to levels inhospitable for human beings—a geophysiological "fever."

The Stuff of Life

When German chemist Friedrich Wöhler (1800-1882) first, accidentally, produced crystals of urea by heating ammonium cyanate, he could not accept that he had made from scratch a compound so clearly associated with living beings. Urea, after all, is the carbon-nitrogenous waste produced in animal urine. And in Wöhler's day, organic beings were believed to consist of a strange and wonderful "organic matter" that was present in life—and nowhere else. Since then, dozens of carbon-rich compounds, such as formic acid, ethylene, and hydrogen cyanide, have been found not just in life but in interstellar space. The equivalent of an estimated 10 quintillion (10,000,000,000,000,000,000) fifths of whiskey, in the form of the nine-atom molecule CH_3CH_2OH (ethyl alcohol), exists in one interstellar cloud in the constellation of Orion alone.

Though adulterated with other compounds, we, like all living matter, are mostly water—that is, hydrogen and oxygen. Hydrogen forms, by mass, seventy-five percent of the atoms in the cosmos. It is the same element which, under intense gravitational pressure, becomes helium in the nuclear fusion reaction that makes our sun shine. Far older and bigger stars went out with a bang, as supernovas, and thereby created carbon, oxygen, nitrogen, and the other heavier elements. Life is made from such star stuff. In the universe life may be rare or even unique. But the stuff of which it is made is readily available.

More and more inert matter, over time, has literally come to life. Minerals of the sea are now incorporated into living creatures for protection or support in the form of integument, shell, bone. Our own skeletons are built from calcium phosphate, a sea salt that was initially a nuisance or a hazard for our remote ancestors, marine protist cells which eventually found ways to cleanse their tissues by putting such minerals to use. The kinds as well as the mass of chemical elements in living bodies have increased through evolutionary time. Whereas structural compounds made of hydrogen, oxygen, sulfur, phosphorus, nitrogen, and carbon are required by all cells and have been essential to life since its inception, those made of silicon and calcium are relative newcomers.

Heinz Lowenstam (1913-1993), a Silesian-born geologist and refugee from Nazi Germany, cataloged the minerals produced in the hard parts of animals. In Lowenstam's youth, the only hard substances thought to be produced by living tissues were the calcium phosphate of our own bones and teeth, the calcium carbonate of mollusk shells, and the silicon dioxide of unusual structures such as the spicules of sponge. [PLATES 8 AND 9, *following page*] Lowenstam and his colleagues went on to discover many other minerals produced by life. The list of hard substances made by live cells, including unexpectedly beautiful crystals, now surpasses fifty. [CHART C, *following page*]

Life had been reusing hard materials and shaping solid wastes long before the appearance of technological humans. Bacteria came together to form protoctists that in turn could mine and use calcium, silica, and iron from the world's seas. Protoctists evolved into animals with shells and bones. Animals, individually or in concert, engineered inert materials into tunnels, nests, hives, dams, and the like. Even some plants incorporate minerals. The silica-laced bodies of "scouring rushes," for example, may serve as good pot scrubbers for campers, but they have probably evolved to deter herbivores. The calcium oxalate crystals of *Dieffenbachia* are hurtled from the leaf cells toward unwary, hungry victims.

The propensity to "engineer" environments is ancient. Today people make over the global environment. Clothed and bespectacled inside an automobile, connected by phone wires and radio

8
Oxalic acid crystal taken
from a sea squirt renal sac,
an organ thought to be a
ductless kidney. *Nephro-
myces*, a protist probably
associated with symbiotic
bacteria, apparently forms
the crystals from the animal's uric acid and calcium
oxalate. Over 50 such minerals are now known to be
produced in living cells.

9
Sea squirts. Phylum:
Chordata. Kingdom:
Animalia. Some of these
sessile sea animals (also
called didemnids or sea
lemons) produce calcium
oxalate crystals such as
that shown in plate 8.

Chart C

Contrary to popular belief minerals and animals do not belong to separate kingdoms. Many minerals are produced in and by life, sometimes in crystalline form. One of the most common minerals, calcium carbonate, is formed by living marine animals as shells. Another compound, calcium phosphate, is precipitated by cells of our bones. As this table shows all five kingdoms of organisms have members which produce minerals. This list represents only a sample of the over fifty minerals now known to be produced by living cells.

Mineral	Bacteria	Protoctista	Fungi	Animals	Plants
CALCIUM					
Calcium carbonate ($CaCO_3$; aragonite, calcite, vaterite)	SHEATH AND OTHER EXTRACELLULAR PRECIPITATE	AMOEBA AND FORAMINIFERAN SHELLS	EXTRACELLULAR PRECIPITATES MUSHROOMS	CORALS MOLLUSK SHELLS ECHINODERM SKELETONS CALCAREOUS SPONGES SOME KIDNEY STONES	EXTRACELLULAR PRECIPITATES
Calcium phosphate ($CaPO_4$)			EXTRACELLULAR PRECIPITATES MUSHROOMS	BRACHIOPOD "LAMP SHELLS" VERTEBRATE TEETH AND BONES SOME KIDNEY STONES	
Calcium oxalate (CaC_2O_4)				MOST KIDNEY STONES	*DIEFFENBACHIA*, A FLOWERING PLANT
SILICON					
Silica (SiO_2)	PRECIPITATES	DIATOM AND RADIOLARIAN SHELLS MASTIGOTE ALGAE SCALES		GLASS-SPONGE SPICULES	GRASS PHYTOLITHS HORSE-TAIL STEMS
IRON					
Magnetite (Fe_3O_4)	MAGNETOSOMES			ARTHROPODS MOLLUSKS VERTEBRATES	
Greigite (Fe_3S_4)	MAGNETOSOMES				
Siderite ($FeCO_3$)	EXTRACELLULAR PRECIPITATES				
Vivianite ($Fe_3[PO_4]_2 \cdot 8H_2O$)	EXTRACELLULAR PRECIPITATES				
Goethite ($xxFeO \cdot OH$)	EXTRACELLULAR PRECIPITATES		EXTRACELLULAR PRECIPITATES	CHITONS	
Lepidocrocite ($xxFeO \cdot OH$)	EXTRACELLULAR PRECIPITATES		EXTRACELLULAR PRECIPITATES MUSHROOMS	CHITONS	
Ferrihydrite ($5Fe_2O_3 \cdot 9H_2O$)				MOLLUSKS	FLOWERING PLANTS
MANGANESE					
Manganese dioxide (MnO_2)		INTRACELLULAR OR EXTRACELLULAR PRECIPITATES AROUND SPORES			
BARIUM					
Barium sulfate ($BaSO_4$)		ALGAL-PLASTID GRAVITY SENSORS MARINE PROTIST SKELETONS		SENSE ORGANS: STATOLITHS (OTOLITHS)	
STRONTIUM					
Strontium sulfate ($SrSO_4$)		MARINE PROTIST SHELLS		MOLLUSK SHELLS	

waves to modems, cellular phones, and bank machines, supplied with electricity, plumbing, and other utilities, we are transforming ourselves from individuals into specialized parts of a global more-than-human being. This metahuman being is inextricably bound to the much older biosphere, from which it arose. Metals and plastics represent the newest realm of matter "coming to life."

Mind in Nature

The biological self incorporates not only food, water, and air—its physical requirements—but facts, experiences, and sense impressions, which may become memories. All living beings, not just animals but plants and microorganisms, perceive. To survive, an organic being must perceive—it must seek, or at least recognize, food and avoid environmental danger.

A living being need not be conscious to perceive. But consider: most of our own daily activities (breathing, digesting, even turning a page or driving a car) are performed largely or even wholly unconsciously. From the viewpoint of the evolutionary biologist, it is reasonable to assume that the sensitive, embodied actions of plants and bacteria are part of the same continuum of perception and action that culminates in our own most revered mental attributes. "Mind" may be the result of interacting cells.

Mind is fully an evolutionary phenomenon. Hundreds of millions of years before organic beings verbalized life, they recognized it. Discerning what could kill them, what they could eat, and what they could mate, roughly in that order, were crucial to animal survival. One U.S. Supreme Court justice avowed that while he might not be able to define obscenity he surely could recognize it when he saw it. We all have a similar ability with life. Life has been recognizing itself long before any biology textbooks were written.

Survival-based psychological tendencies infiltrate the pristine realm of science. Pattern recognition was such a useful trait for our ancestors that, even if occasionally wrong, the *Aha!* feeling of discovery would have been reinforced. Aesthetic judgments of elegance and beauty, often cited in the preference for certain equations over others in physics, show that scientific correctness can also be intuitive. What we know, what we are capable of knowing and seeing, has been shaped by our evolution as surviving creatures. Even foolish and outlandish notions would have been retained and reinforced if they in any way aided our ancestors' survival.

Neuroscientists have traced subjective feelings of pleasure to endorphins and enkephalins, two groups of neuropeptides produced by the brain. The pleasure associated with seeing beauty, including scientific "truth," may have come about during the course of evolution, just as love and biophilia—the pleasure we take in the company of other living creatures—provoke us to seek out mates and the natural environments that have been most conducive to our survival. If we did not fear death, we might be too quick to kill ourselves when troubled or inconvenienced and thus perish as a species. Belief in life's importance may not be a reflection of reality, then, but an evolutionarily reinforced fantasy that prejudices believers to do what is necessary, bear whatever burdens, to survive.

We all inherit a shared perspective bequeathed by our ancestors. The physicists' hope of solving an essential set of equations for all time and the cosmos may be but the gleam of a receding mirage. In the end, as Charles Peirce and William James recognized, there may be no better measure of "truth" than that which works—that which helps us survive.

Mind and body, perceiving and living, are equally self-referring, self-reflexive processes already present in the earliest bacteria. Mind, as well as body, stems from autopoiesis. And in sufficiently expressive humans the process of autopoiesis underlying living organization makes itself manifest even outside the body. Abstract expressionist painter Willem de Kooning wrote:

> If you write down a sentence and you don't like it, but that's what you wanted to say, you say it again in another way. Once you start doing it and you find how difficult it is, you get interested. You have it, then you lose it again, and then you get it again. You have to change to stay the same.[10]

Changing to stay the same is the essence of autopoiesis. It applies to the biosphere as well as the cell. Applied to species, it leads to evolution.

So, what is life?

It is a material process, sifting and surfing over matter like a strange, slow wave. It is a controlled, artistic chaos, a set of chemical reactions so staggeringly complex that more than eighty million years ago it produced the mammalian brain that now, in human form, composes love letters and uses silicon computers to calculate the temperature of matter at the origin of the universe. Life, moreover, appears to be on the verge of perceiving for the first time its strange but true place in an inexorably evolving cosmos.

Life, a local phenomenon of Earth's surface, can in fact be understood only in its cosmic milieu. It formed itself out of star stuff, shortly after Earth 4,600 million years ago congealed from a remnant of a supernova explosion. Life may end in a mere hundred million years when, embattled by dwindling atmospheric resources and increased heat from the sun, systems of global temperature regulation finally fail.[11] Or life, enclosed in ecological systems, may escape and watch from safe harbor as the sun, exhausting its hydrogen, explodes into a red giant, boiling off Earth's oceans, five thousand million years from now. ∎

2
Lost Souls

Ay, but to die, and go we know not where;
To lie in cold obstruction and to rot.

—**William Shakespeare,** *Measure for Measure*

Love that endures for a breath:
Night, the shadow of light,
And life, the shadow of death.

—**Algernon Swinburne,** *Atalanta in Calydon*

Death: The Great Perplexer

The scientific mystery of life in a near-lifeless, mechanical universe mirrors the enigma of death in a fully living, animistic one. Our ancestors inhabited a world where warm, moving bodies would regularly stop, grow cold, and decay. As puzzling as life is for us, so was death for them. But we moderns still feel the influence of ancient solutions to the death puzzle.

Until the seventeenth century the sun and moon did not move according to Newtonian principles; these celestial bodies often were animated by spirits within them. The whistling of the wind, the changing phases of the moon, the twinkling, turning stars—these eternal, celestial bodies moved as they

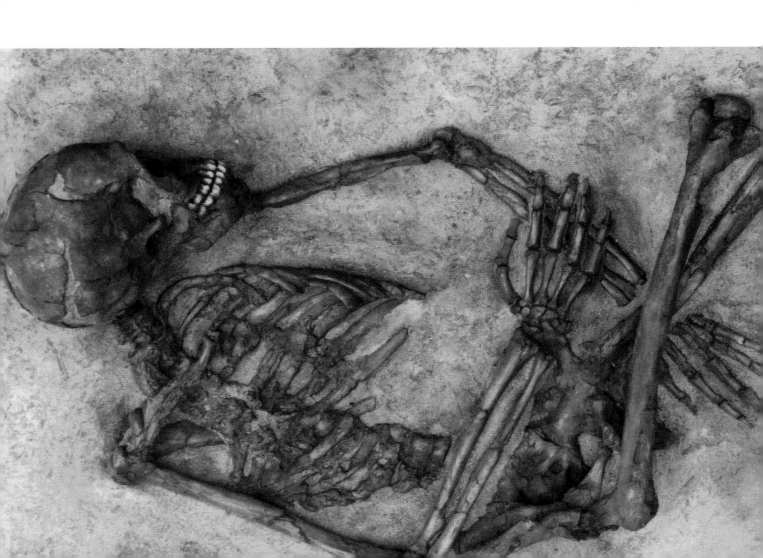

investigation, but in the "fine print" is found the exception: the conscious human soul—which, in Descartes' time was unquestionably made in God's image. Moreover, the Cartesian permit still contains in the fine print this assumption: the universe is mechanical and set up according to immutable laws. Neither the exception nor the assumption is science. At the very heart of the Cartesian philosophy are thus metaphysical presuppositions, springing from the culture that gave rise to science.

Ultimately—in our very abbreviated story—the Cartesian license proves to be a kind of forgery. After three centuries of implicit renewal, the license is still accepted even though the fine print, erased or ignored, is no longer visible at any magnification. Yet this fine print was not incidental. It was the raison d'être, the rational basis authorizing scientists following the spirit of Descartes to proceed with their work and to receive the blessings of society, if not always the Church. The Cartesian view of cosmos as machine is at the very root of the practice of science.

Entering the Forbidden Realm

While Descartes cogitated, Europe remained under the rule of royalty. The King and the Lord, representing the power and order of God, reigned supreme. But science soon entered the forbidden realm of humankind, the one place it was not supposed to go. Scientific revelation of mechanism, part of the new audacity of inquiry, helped unsettle European monarchy. If the universe made by God is a giant automaton that works itself, why should people obey any King or Lord whose power, God-given in the feudal system of medieval Christianity, no longer derived from heavenly decree? The high-born Frenchman Donatien Alphonse François Sade keenly felt the vanishing basis for morality. If Nature was a self-perpetuating machine and no longer a purveyor of divine authority, then it did

not matter what he, as the infamous marquis de Sade, did or wrote.

In 1776 the British colonists in North America broke free from transatlantic rule. Independence from the burdens of taxes and royalty was proclaimed. In 1789 the French Revolution deposed the king and stripped the lords and ladies of their powers. Irreverent Voltaire claimed that if God did not exist it would be necessary to invent him. (A century later German philosopher Friedrich Nietzsche would declare God dead.) England, too, was struck by the revolutionary spirit of the time, but in moderation. Retaining their king and queen, the English perceived themselves a bastion of order in a world gone mad.

Enter Charles Darwin. In 1859 his *Origin of Species* was published, announcing to the world the scientifically derived inference that man had not been created by God, but had evolved from mere animals through "natural selection." Darwin's later books, *Descent of Man* (1871) and *Expression of Emotions* (1872), explored the then-startling thesis that humans and apes evolved from ancient apes. Darwin documented, without any explicit anti-Christian statement, that neither humans nor ancestral apes were created by God. The Great Chain of Being—the line of holiness coming down from God through spiritual angels to humankind and thence to the rest of mechanical creation—was turned topsy-turvy. The cosmic apple cart was upset. No longer, Darwin insinuated, was man excluded from connection with nature. Even the perceiving mind, describing itself, evolved from mechanical laws of random variation and natural selection. Materialism was victorious. As in some maudlin Disney animation, the last sparkle of fairy stuff disappeared.

Western thought thus suffered a metaphysical reversal. Once, before the exploits of Bruno and Galileo, Descartes and Newton, and Darwin,

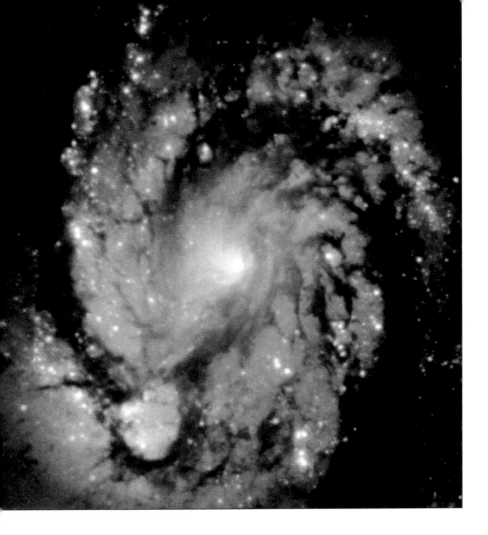

11
The so-called "Grand Design" spiral galaxy, M-100, as taken from the Hubble Space Telescope on December 31, 1993. Our ancestors inhabited a world of living spirits from vengeful storm clouds to serenely shining stars; today science pictures the universe as an inanimate morass of chemical elements and compounds, some of which, under certain conditions, conjoin to form the self-repeating structures of life.

everything had been alive, except for the natural magic trick of death; now in the scientific-mechanistic world everything was inanimate, dead, except for the scientific puzzle of life. [PLATE 11]

We all are interested in life because we know it from the inside as something more than mechanical, automatic, determined responses to preordained stimuli. We think, act, choose. We—and it would be a conceit to exclude other organic beings—are not Newtonian machines.

Moreover, we are not objective outsiders. In physics, Werner Heisenberg's uncertainty principle limits what is measurable. In mathematics, Kurt Gödel's incompleteness theorem warns that every mathematical system, if complete, cannot be consistent and, if consistent, cannot be complete, since to define it axioms are needed from outside the system. Such scientific uncertainty also impedes any search to define life. On the one hand, a final definition of life by life may be like kissing your elbow or rolling your eyes to see your own optic nerve: impossible. On the other hand, enlightened by a knowledge of history and science's astounding success at investigating what life is, we seem closer than ever to a deeper understanding of life in its cosmic and cultural context.

In the flush of this exhilarating material success scientists tend to gloss the distinction between life and nonlife, pointing to the chemical continuities. Life-as-a-whole is like other vast subjects: nationalism, culture, politics, or anything else not easily defined, manipulated, or described. Even biologists may be snide, dismissing relevant discussion as "just philosophy." But science, like anything else, has a context. And that context is partly metaphysics, great, often-unstated categories of thought, perhaps cultural, perhaps inherited (the distinction is itself metaphysical!) that go beyond science proper. No one escapes metaphysics; to understand life, as science, it is necessary to understand its cultural context.

"Metaphysics," introduced by Hellenistic scholars and referring to certain untitled texts by Aristotle, comes from the Greek *ta meta ta physika biblia*, which literally means "the books after ('meta') the books on nature." The original use of the prefix "meta," by early editors such as Andronicus of Rhodes, may not have referred to any transcendental interpretation of ultimate reality, but only to the mundane position of the book on the table where "Metaphysics" was stacked on top of "Physics." Beginning with the work of Immanuel Kant, metaphysics has come to refer to speculations on questions not answerable by direct observation or experiment. Metaphysics, as a web of ideas in which we are caught, need not give rise to futility. It is fascinating to try and tease apart the strands of the culturally inherited, linguistically reinforced concepts that guide even our most seemingly original thoughts. An explanation of metaphysics may not lead to absolute truth but it certainly shouldn't be anathema to open, scientific minds.

Cosmic Wiggles

"A living body," wrote Alan Watts, "is not a fixed thing but a flowing event." Watts, the Anglo-American popularizer of eastern philosophy, drew from science, as well, in his quest for the meaning of life. He likened life to "a flame or a whirlpool":

> the shape alone is stable. The substance is a stream of energy going in at one end and out at the other. Life's purpose to maintain and perpetuate itself is understandable as a physico-chemical phenomenon studied by the science of thermodynamics. We are temporarily identifiable wiggles in a stream that enters us in the form of light, heat, air, water, milk. . . . It goes out as gas and excrement—and also as semen, babies, talk, politics, war, poetry and music."[6]

Thermodynamic systems lose heat to the universe as they convert energy from one form to another. Living matter frees itself from ordinary matter only by perpetually basking in the sun. Confronted with dissolution and destruction, life suffers a permanent death threat. Life is not merely matter, but matter energized, matter organized, matter with a glorious and peculiar built-in history. Life as matter with needs inseparable from its history must maintain and perpetuate itself, swim or sink. The most glorious organic being may indeed be nothing but "temporarily identifiable wiggles," but for millions of years as life has been racing away from disorder autopoietic beings have concerned themselves with themselves, becoming ever more sensitive, ever more future oriented, and ever more focused on what might bring harm to the delicate wave of their matter-surfing form. From a thermodynamic, autopoietic perspective, the basest act of reproduction and the most elegant aesthetic appreciation derive from a common source and ultimately serve the same purpose: to preserve vivified matter in the face of adversity and a universal tendency toward disorder.

Dutch-Jewish philosopher Baruch Spinoza (1632-1677) portrayed matter and energy as the fundamental nature of a universe which was itself alive. The great German writer and naturalist Johann Wolfgang von Goethe (1749-1832), author of *Faust*, argued for a poetic biology. He thought matter does not operate without spirit, nor does spirit exist without matter. Although he was pre-Darwinian and his theories are now obsolete,

Goethe wrote ably on science. In one passage he plucks from human activity what might be called its autopoietic essence:

> Why are the people thus busily moving? For food they are seeking,
>
> Children they fain would beget, feeding them well as they can.
>
> Traveler, mark this well, and, when thou art home, do thou likewise!
>
> More can no mortal effect, work with what ardor he will.[7]

The German biologist Ernst Haeckel (1834-1919), inventor of the word "ecology," promoted the idea that the activity of the human psyche is an offshoot of physiology: "We hold with Goethe that matter cannot exist...without spirit.... We adhere firmly to the pure, unequivocal monism of Spinoza: Matter, or infinitely extended substance, and Spirit (or Energy), or sensitive and thinking substance, are the two fundamental attributes, or principal properties, of the all-embracing essence of the world, the universal substance."[8]

The Meaning of Evolution

Ernst Haeckel was Darwin's translator and greatest advocate in the German tongue, but he pushed Darwinism further than its inventor had been willing to carry it. The soul, Haeckel claimed, resided in the cell, immortality was a metaphysical sham, life had no purpose other than itself, and beings were not spiritual but material in nature. "Humanity," he declared, "is but a transitory phase of evolution of an eternal substance, a particular phenomenal form of matter and energy, the true proportion of which we soon perceive when we set it on the background of infinite space and eternal time."[9]

Such views infuriated traditional religious sensibilities, including those of Alfred Russel Wallace (1823-1913). An English naturalist, Wallace developed his own theory of evolution by natural selection that was uncannily similar to that of Darwin. Darwin's and Wallace's short papers on natural selection were published together in the same issue of the *Journal of the Proceedings of the Linnean Society of London, Zoology*. Wallace, who frequented seances, reviled Haeckel's notion of matter as eternal and alive, and he rejected Haeckel's denial of a spirit world. He sneered that *The Riddle of the Universe*, the title of one of Haeckel's most influential and popular books, had not been solved, least of all by Haeckel.

Even before Darwin, German philosopher Immanuel Kant (1724-1804) noted that skeletal and other similarities pointed to blood ties, a common parentage for all life. Kant ceded that all life could have arisen through some mechanical process similar to that by which nature produces crystals, but he judged it would be absurd to hope for "a Newton" who could make comprehensible even the growth of a single blade of grass by mechanical theory alone. Haeckel proposed Darwin as the very "Newton" Kant had believed impossible.

By projecting Earth history millions of years beyond the six thousand years allotted in the "Book of Genesis," James Hutton (1726-1797) founded modern geology. Hutton, son of a Scottish merchant, distinguished rocks laid down as sediment from those brought forth in molten form through volcanos. He observed erosion by wind and water and deduced the production of rainfall from cooling air masses that could no longer contain their moisture. Older sediments were deposited prior to more recent ones. Hutton's "law of superposition" led to Charles Lyell's (1799-1875) statement of the "law of uniformitarianism," the suggestion that only those geological forces observable in the present need be invoked to account for structures made and sediments accumulated in the past. But Hutton's extrapolation that Earth must be very old was controversial. Conservative England, threatened by the wild and godless French Revolution, was not ready to accept an Earth older than that which could be ascertained by summing up all the begats mentioned in *The Bible*.

Nonetheless, Scottish geologist Charles Lyell approved Hutton and argued that time was far vaster than previously thought in his multivolume book, *The Principles of Geology*—which did for that field what Darwin's opus later did for zoology and botany. Lyell was also far ahead of his time in taking a global ecological perspective reminiscent of Gaia theory today; he called attention to "the powers of vitality on the state of the earth's surface."[10] Darwin read Lyell during his voyage on the *Beagle* and adopted the Lyellian worldview. Decades later Lyell, in turn, embraced the Darwinian worldview. In 1863 he published *The Antiquity of Man*, which suggests, before Darwin had made the extension, that evolution applied to all humankind.

Meanwhile on the Continent, Berlin naturalist Christian Gottfried Ehrenberg (1795-1876) was putting the life back into biology. Returning from an ill-fated expedition to Egypt, of which he was perhaps the sole survivor, he focused on the transition between life and nonlife. In the expedition to Egypt (1820) and a later one to Siberia (1829) Ehrenberg documented the unseen world of microbes that fertilize the oceans and soils. Through his journeys Ehrenberg came to know Friedrich Wilhelm Heinrich Alexander, the Baron von Humboldt (1769-1859). Humboldt, widely regarded as the greatest German naturalist of his time, had collected more than sixty thousand plant specimens during his travels around the world. He had visited American president Thomas Jefferson and was described as a scientific "Napoleon." In his seventies

Humboldt began to compile *Kosmos*, his grand attempt to map and explain the entire universe. "Certainly," wrote Isaac Asimov, "no man before him, with so active a mind, had seen so much of the world, and no man before him was so well equipped to write such a book.... It was a florid production, rather overblown, but it is one of the remarkable books in scientific history and was the first reasonably accurate encyclopedia of geography and geology."[11]

In *Kosmos*, Humboldt shares Ehrenberg's discovery of life's global sweep. "The universality of life is so profusely distributed," waxes Humboldt,

> that the smaller Infusoria [ciliates and other protists] live as parasites on the larger, and are themselves inhabited by others.... The strong and beneficial influence exercised on the feeling of mankind by the consideration of the diffusion of life throughout the realms of nature is common to every zone, but the impression thus produced is most powerful in the equatorial regions, in the land of palms, bamboos, and arborescent ferns, where the ground rises from the shore of seas rich in mollusca and corals to the limits of perpetual snow. The local distribution of plants embraces almost all heights and depths. Organic forms not only descend into the interior of the earth, where the industry of the miner has laid open extensive excavations and sprung deep shafts, but I have also found snow-white stalactitic columns encircled by the delicate web of an *Usnea* [old man's beard lichen], in caves where meteoric water could alone penetrate through fissures.... [Organisms flourish on the summits of the] Andes, at an elevation of more than 15,000 feet. Thermal springs contain small insects (*Hydroporus thermalis*), *Gallionellae* [iron bacterial masses], *Oscillatoria*, and *Confervae* [an old name for a

miscellany of green algae], while their waters bathe the root-fibers of phanerogamic [cone- and flower-bearing] plants.[12]

Humboldt died the same year Darwin published *The Origin of Species.*

Until very recently, with publication of the work of Schrödinger's legacy, observations made by Humboldt and Ehrenberg on the microbial world and many other late nineteenth century discoveries were not brought together in an evolutionary context. The fertilization of sperm by egg (embryo formation), inheritance factors of garden peas (Mendelian genetics), mucoid substances in the pus of soldier's wounds (nucleic acids, DNA and RNA), and visualization of chromosomes were some of the revelations made last century which, in geneticist Theodosius Dobzhansky's words, only "make sense…in the light of evolution."[13]

Although theories of evolution had been in the air for a half century and more, Darwin's methodical purposefulness, his diplomacy of prose, and his presentation as an Englishman of a mechanical theory during a time when Isaac Newton's theory of gravity was the last word in science all helped make the appearance of his book an epic event. As one woman of society wryly remarked on hearing the news of her less-than-noble origins, "Let us hope it is not true. But if it is, let us hope it does not become generally known."

Since *The Origin,* the idea of evolution has become increasingly accepted — overwhelmingly by scientists and respectably by the public (particularly the educated public). But it has also been abused. For example, in a popular illustration Haeckel depicted the summit of evolution as a nude but demure Germanic woman at the top of his evolutionary tree. Haeckel's error was not so much in his Germanic bias (or his choice of the female sex) but in his choice of any human at all. This is because all extant species are equally evolved. All

living beings, from bacterial speck to congressional committee member, evolved from the ancient common ancestor which evolved autopoiesis and thus became the first living cell. The fact of survival itself proves "superiority," as all are descended from the same metabolizing Ur-form. The gentle living explosion, in a circuitous 4,000-million-year path to the present, has produced us all. In a sense then, the Vedic intuition that individual awareness is illusory and that each of us belongs to a single primal ground — Brahman — may be accurate: we share a common heritage, not only of chemistry but of consciousness, of the need to survive in a cosmos whose matter we share but which is itself indifferent to our living and self-concern.

Vernadsky's Biosphere

Given the limited legacy of metaphysical dualism (mind/body, spirit/matter, life/nonlife), it may not be surprising that two of the most profound rethinkers of life and its environment in this century share a biospheric perspective yet have diametrically opposed views. But whereas Russian scientist Vladimir Ivanovich Vernadsky (1863-1945) described organisms as he would minerals — calling them "living matter" — English scientist James E. Lovelock describes Earth's surface, including rocks and air, as alive.

Vernadsky portrayed living matter as a geological force — indeed, the greatest of all geological forces. Life moves and transforms matter across oceans and continents. Life, as flying phosphorus-rich gulls, racing schools of mackerel, and sediment-churning polychaete worms, moves and chemically trans- forms the planet's surface. Moreover, life is now known to be largely responsible for the unusual character of Earth's oxygen-rich and carbon diox- ide-poor atmosphere.

Like Ehrenberg and Humboldt before him, Ver- nadsky showed what he called the "everywhereness of life" — living matter's almost total penetration into, and consequent involvement in, seemingly inanimate processes of rock, water, and wind. Others spoke of an animal, vegetable, and mineral kingdom; Vernadsky analyzed geological phenomena without preconceived notions of what was and was not alive. Perceiving life not as life but as "living matter," he was free to broaden its study beyond that of biology or any other traditional discipline. What struck him most was that the material of Earth's crust has been packaged into myriad moving beings whose reproduction and growth build and break down matter on a global scale. People, for example, redistribute and concentrate oxygen, hydrogen, nitrogen, carbon, sulfur, phosphorus, and other elements of Earth's crust into two-legged, upright forms that have an amazing propensity to wander across, dig into, and in countless other ways alter Earth's surface. We are walking, talking minerals.

Vernadsky contrasted gravity, which pulls material vertically toward the center of Earth, with life—growing, running, swimming, and flying. Life, challenging gravity, moves matter horizontally across the surface. Vernadsky detailed the structure and distribution of aluminosilicates in Earth's crust and was the first to recognize the importance to geological change of heat released from radioactivity.

But even a resolute materialist like Vernadsky found a place for mind. In Vernadsky's view a special thinking layer of organized matter growing and changing Earth's surface is associated with humans and technology. To describe it, he adopted the term *noosphere*, from Greek *noos*, mind. The term had been coined by Edouard Le Roy, philosopher Henri Bergson's successor at the College de France. Vernadsky and Le Roy met in Paris for intellectual discussions in the 1920s, along with Pierre Teilhard de Chardin, the French paleontologist and Jesuit priest whose writings would later bring the idea of noosphere—a conscious layer of life—to a wide audience. Teilhard's and Vernadsky's use of the term noosphere, like their slants on evolution in general, differed. For Teilhard the noosphere was the "human" planetary layer forming "outside and above the biosphere," while for Vernadsky the noosphere referred to humanity and technology as an integral part of the planetary biosphere.

Vernadsky distinguished himself from other theorizers by his staunch refusal to erect a special category for life. In retrospect we can see the value of his stance; because life has indeed become a category, theorists of life have managed to reify—to make a thing out of—something that is not a thing at all. Vernadsky's referring to life as "living matter" was no mere rhetorical ploy. In one deft verbal stroke Vernadsky cut loose centuries of mystic clutter attached to the word "life." Making every attempt to consider life part of other physical processes, his use of the gerund "living" stressed that life was less a thing and more a happening, a process. Organisms for Vernadsky are special, distributed forms of the common mineral, water. Animated water, life in all its wetness, displays a power of movement exceeding that of limestone, silicate and even air. It shapes Earth's surface. Emphasizing the continuity of watery life and rocks, such as that evident in coal or fossil limestone reefs, Vernadsky noted how these apparently inert strata are "traces of bygone biospheres."[14]

Austrian geologist Edward Seuss had coined the word "biosphere," but Vernadsky brought it into currency. Just as the sphere of rock is a lithosphere, and that of air an atmosphere, so the sphere where life exists is a "biosphere." In his 1926 book, *The Biosphere*, Vernadsky showed how Earth's surface was an ordered transformation of the energies of the sun. "The biosphere," wrote Vernadsky, "is as much, or even more, the creation of the Sun as it is a manifestation of Earthly processes. Ancient religious intuitions that regarded Earth's creatures, especially humans, as 'children of the Sun' were much nearer

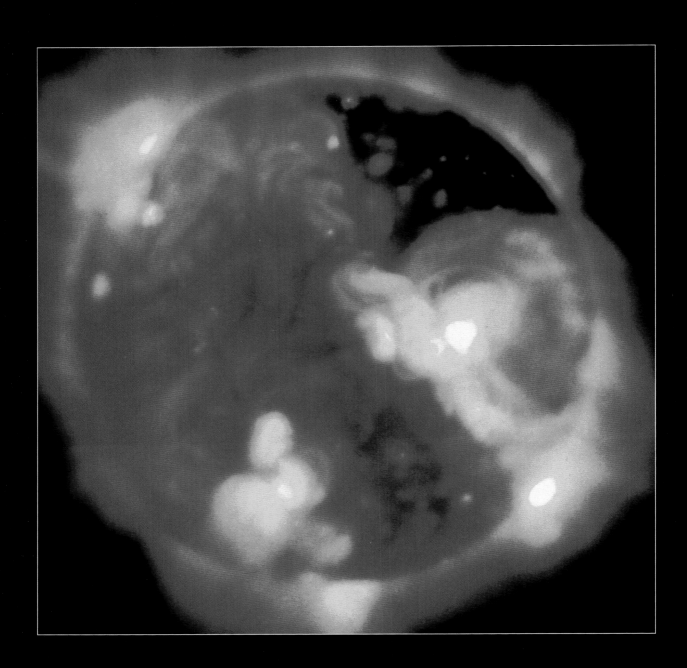

12
NASA x-ray photograph of the Sun. According to Russian scientist Vladimir Vernadsky (1863-1945) life on Earth is a material system in which stellar energies become transformed into living ones. Life is not only a global but a planetary/solar system phenomenon.

the truth than those that looked upon them...as a mere ephemeral creation, a blind and accidental product of matter and Earth forces...We may regard living matter in its entirety, then as the peculiar and unique domain for the accumulation and transformation of the luminous energy of the Sun."[15]

Remarkably, Vernadsky dismantled the rigid boundary between living organisms and a nonliving environment, depicting life globally before a single satellite had returned photographs of Earth from orbit. Indeed, Vernadsky did for space what Darwin had done for time: as Darwin showed all life descended from a remote ancestor, so Vernadsky showed all life inhabited a materially unified place, the biosphere. Life was a single entity, transforming to earthly matter the cosmic energies of the sun. [PLATE 12] Vernadsky portrayed life as a global phenomenon in which the sun's energy was transformed. Emphasizing photosynthetic growth of red and green bacteria, algae, and plants, he saw these expressions of living matter as the "green fire" whose expansion, fed by the sun, pressured other beings into becoming more complex and more dispersed.

Vernadsky set forth two laws. Over time, he claimed, more and more chemical elements became involved in the cycles of life. Second, the rate of migration of atoms in the environment has increased with time. A flock of migrating geese was to Vernadsky a biospheric transport system for nitrogen. Locust swarms, recorded in *The Bible*, attested to massive changes in the distribution of carbon, phosphorus, sulfur, and other biologically important chemicals two thousand years ago. As dams, factories, mines, machine construction, utilities, trains, planes, global communications, and entertainment systems have appeared, more chemical elements than ever have become organized into functioning parts of autopoietic systems. Technology, from a Vernadskian perspective, is very much a part of nature. The former calf muscle severed into brochette cubes and the pine tree trunk

into lumber pass through the hands of workers and the chutes of machines to emerge transformed into shish kebob and floor. The plastics and metals incorporated in industry belong to an ancient process of life co-opting new materials for a surface geological flow that becomes ever more rapid. And, with the fleeting synthesis in physicists' laboratories of radioactive isotopes, the noosphere begins to direct and organize atoms that have never before existed on Earth.

Lovelock's Gaia

As Vernadsky disrupted the mind/matter split through a consideration of living matter on a global scale, so James E. Lovelock upsets metaphysical dualism by an opposite stratagem—considering Earth alive. Vernadsky examined life as matter within a receptive political and cultural climate—the official atheism of the former Soviet Union, aided by science's approval of materialism. By contrast, Lovelock—portraying the self-regulating biosphere, a huge and oddly spherical living body he calls "Gaia"— has been hampered by the subtle ideology of mechanism that pervades the scientific community. This means that Lovelock must not only show that Earth maintains itself as a living body, he must also overcome the prejudice that to call this "thing" alive is not science but poetic personification. Given these tensions, it is a testament to this world-class atmospheric chemist's ingenuity that his theory is taken as seriously as it is by active scientists.[16]

Atmospheric, astronomical, and oceanographic evidence attest that life manifests itself on a planetary scale. The steadiness of mean planetary temperature for the past three thousand million years, the 700-million-year maintenance of Earth's reactive atmosphere between high-oxygen levels of combustibility and low-oxygen levels of asphyxiation, and the apparently continuous removal of hazardous salts from oceans—all these

13
Emiliana huxleyi, a coccolithophorid. Phylum: Haptophyta. Kingdom: Protoctista. This coccolithophorid, a calcium-precipitating alga, is covered with button-like scales. These protists, each only 20 millionths of a meter in diameter, produce dimethyl sulfide, a gas of global significance involved in cloud cover over the ocean.

14
A 50 kilometer-wide bloom of coccolithophorid scales extends 200 kilometers along the coast of Scotland as seen in a satellite image. These tiny beings, implicated in the production of cloud cover and constructing their shells from the greenhouse gas carbon dioxide, may play a role in global climate control. They are a striking example of the "fractal" magnification of microbial into biospheric behavior.

point to mammal-like purposefulness in the organization of life as a whole. [PLATES 13 and 14]

This purposefulness, central to scientific Gaia theory, is a major sticking point for traditional biologists. How can a planet behave in a purposeful manner to maintain environmental conditions favorable to its living constituents? In mechanistic biology, complex self-regulation only evolves from natural selection that weeds out more poorly self-regulating individuals. This logic is flawed, however. According to it no original, self-maintaining cell could have ever evolved, because "purposeful," self-regulating behavior simply cannot arise in a population with only one member. A strict reading of Darwinism denies evolutionary capabilities to a population of one.

Plausible within the bounds of Darwinism or not, both planet, isolated by space, and cell, isolated by semipermeable membrane, are solar energy-requiring systems, continuous through time and space, that display self-maintaining behavior. The "purposefulness" of Gaian self-maintenance derives from the living behavior of myriad organisms, mostly microbes, whose ubiquity Ehrenberg and Humboldt first established. Planetary physiology, far from having been produced *ex nihilo*, or by an outside God, is the holarchic outcome of ordinary living beings. It is the autopoiesis of the cell writ large.

Life cannot be understood ignoring the sentient observer. If not for mind, no one would care that life is a certain kind of sunlight-energized cosmic debris. But it is, and we do. To best understand life we need to see the long and winding road from animism, through dualism, to the limitations of mechanism. Physics, chemistry, and biology are distinct approaches to the same material phenomena. As German geomicrobiologist Wolfgang Krumbein puts it,

> The mineral and microbial mineral cycles as we view them today on the basis of experimental work have been envisaged as the unifying

concept of world and universe, creating the principle of the one living nature of Bruno and Spinoza....The basic approach of Bruno... is still alive and is evidenced in scientific and mathematic terms by non-Euclidian geometry, by the modern field theories and Einstein's relativity and gravity theories, as well as by Lovelock's "Gaia-hypothesis." Giordano Bruno deeply influenced Baruch Spinoza, Leibnitz, Kant, Goethe and Schelling. He still influences unitarian thought in science and philosophy... a textbook on microbial geochemistry...must come back to Bruno's original thoughts of "cyclic developments" rather than "creation and destiny" as revealed in the clerical Christian thoughts of his time which have so severely inhibited the development of science.[17]

Life can be returned to biology without compromising science. Mechanism gave science the authority to examine the realms of heaven and life once considered "off limits." But it also suggested the universe was more deterministic than it is, cutting into our sense of life and wonder. The Epicurean Roman philosopher Lucretius (95-55 B.C.), in his poem *De Rerum Natura* ("On the Nature of Things"), presents an evolutionary view of the universe denying a hereafter and arguing that everything, even the soul and gods, are made of atoms. In the same tradition, Bruno blended matter with energy, finite with infinite, world with God. In the modern era, by not speaking of life at all — but calling it "living matter"—Vernadsky offered us a chance to see life with fresh eyes. And, unlike monolithic Cartesian materialism, the Gaia perspective accommodates the enchantment we feel as living beings dwelling in a living world.

So, what is life?

Life is planetary exuberance, a solar phenomenon. It is the astronomically local transmutation of Earth's air, water, and sun into cells. It is an intricate pattern of growth and death, dispatch and retrenchment, transformation and decay. Life is the single expanding organization connected through Darwinian time to the first bacteria and through Vernadskian space to all citizens of the biosphere. Life as God and music and carbon and energy is a whirling nexus of growing, fusing, and dying beings. It is matter gone wild, capable of choosing its own direction in order to indefinitely forestall the inevitable moment of thermodynamic equilibrium— death. Life is also a question the universe poses to itself in the form of a human being.

What happened to living matter to make it so different? The answer is both scientific and historical. Life is its own inimitable history. From an everyday, uncontentious perspective "you" began in your mother's womb some nine months before whatever your age is. From a deeper, evolutionary perspective, however, "you" began with life's daring genesis—its secession, more than 4,000 million years ago, from the witches brew of the early Earth. In the next chapter we see how this brew, sometimes called the primeval soup, started percolating.

3

Once Upon the Planet

If a dirty undergarment is squeezed…a ferment
drained from the garments and transformed by the
smell of the grain, encrusts the wheat itself with its
own skin and turns it into mice…. And, what is more
remarkable, the mice from corn and undergarments
are neither weanlings or sucklings nor premature
but they jump out fully formed.

— **Jean Baptiste van Helmont**

For the humblest organism, the simplest bacterium,
is already a coalition of enormous numbers of
molecules. It is out of the question for all the pieces to
have been formed independently in the primeval
ocean, to meet by chance one fine day, and suddenly
arrange themselves in such a complex system.

— **François Jacob**

It must be admitted that no one yet
knows how life began.

— **Stanley Miller** and **Leslie Orgel**

Beginnings

On Earth some four thousand million years
ago life generated as matter undertook a different
direction. From the beginning, life satisfied its
autopoietic imperative in a universe obeying
thermodynamic laws. Bound and separated from
the world by a border of its own making, life came
together as oily droplets that increased their order.
[PLATE 15] Other dissipative systems in nature use
energy to increase order, but they last for only short
periods of time. Moreover, a tornado risen on the
plains doesn't go "whoops" as it wanders into a
mountainous landscape that spells its doom; but
even the simplest life form effectively does, actively
responding to its surroundings to preserve and
protect its form.

How matter in a bath of energy (or how energy
in a brew of matter) first accomplished the feat of
life is not known. No molecule by itself reproduces.
Minimal life on Earth today is a system, a minute
membrane-bounded sphere, a bacterial cell, requiring many interacting molecules. Some 2,000 to
5,000 genes make a similar number of proteins. Proteins and DNA mutually produce each other within
the cell membrane that together they fabricate. Bearing a common biochemistry, all life probably dates
to a single, perhaps (but not necessarily) improbable
historical moment. The factors that led matter to
its peculiar "fractionation point" where dissipative
behavior became living behavior need only have
happened once. Enclosed, perhaps even suddenly, by
a membrane and with resources aplenty, the first
living cells could afford to be somewhat aloof from
external reality. Eventually, imperiled by its own
profligacy and by the insensitivity of the substance
from which it seceded — yet upon which it absolutely
depended for sustenance — life was left to its own
devices. As matter ostracized from itself, life had
been abandoned by the world, yet the world had
gone nowhere. There was no going back.

Once begun, reproducing systems proceeded

(Millions of years ago)
4,600   **4,500**   **4,400**   **4,300**

HADEAN EON

Origins Earth-Moon system and other solar system planets.

Oldest rocks dated by radioactivity from meteor (a chondrite from Canyon Diablo crater, Arizona).

Outgassing of volatiles from mantle to atmosphere.
Abundant impact cratering.

Oldest known mineral crystals (present-day Australia).
Possible existence of first continents.

semina, invisible seed. Around the year 1000, Cardinal Pietro Damiani insisted that birds bloom from fruits, and ducks emerge from sea shells. English scholar Alexander Neckam (1157-1217) specified that fir trees, exposed to sea salt, give rise to geese. The Flemish alchemist and physician Jan Baptiste van Helmont (1580-1644) shared his recipe for making mice from dirty underwear.

We moderns may laugh, but the notion of spontaneous generation made such sense at the time that few questioned it. "Since so little is required to make a being," agreed Descartes, "it is certainly not surprising that so many animals, worms, and insects form spontaneously before our eyes in all putrefying substances."[3] Aristotle had taught that the heat of the male seed animated and formed the cooler matter carried in the woman's womb. Lacking sufficient male heat, a woman miscarried or gave birth to a limbless infant. Heat might bypass seed altogether and directly generate worms, bats, snakes, crickets, or other vermin from meat or filth. Alchemists used heat to try to synthesize gold. In a patrilineal, male-dominated Europe women were like potter's kilns in which the act of fathering came to fruition; the female supplied only matter and not the essence of living form. Even Newton suggested that plants might spring forth from the coruscation of cometary tails. Nor did the invention of the microscope sweep away the old idea. Many believed that the "animalcules" Leeuwenhoek had discovered in plant fluids, ditch water, and saliva emerged directly from these fluids, just as veal—left to its own devices—was thought to generate flies.

Ironically, the notion of spontaneous generation was at first threatened as much by the idea of fixed species as it was by countervailing observations.

Species were recorded as fixed categories. The works of Swedish botanist Carolus Linnaeus (1707-1778), the founder of modern taxonomy who gave the name *Homo sapiens* to the human body (but not to the soul), and those of French anatomist Georges Cuvier (1769-1832), who extended the Linnaean classification to fossils, made the notion of spontaneous generation more difficult to accept. For Linnaeus, fixed species were distinct and separate forms created by an omnipotent God. Cuvier believed fossils evidence of past life, in particular of catastrophic floods, at least one of which was recorded in *The Bible.*

It thus came to be believed that an all-powerful God, once and for all, had created all Earth's "creatures." Indeed, Swiss naturalist Charles Bonnet (1720-1793) ruled out spontaneous generation because it was superfluous to his theory of *emboîtement*—that the original female of each species had been created, as in a set of Russian dolls, with the germ cells of all future generations already inside her. Half-blind Bonnet had discovered the all-female reproduction system of certain insects, the parthenogenesis of aphids, a fact which helped him argue against "evolution"—the word he used to refer to the belief of those foolhardy enough to believe in the wanton notion of species transformation.

Spontaneous generation was adhered to even after the Florentine physician and poet Francesco Redi (1626-1697) performed his diligent experiments disproving spontaneous generation. Redi placed a variety of meats—a snake, some fish, and a slice of veal—in sealed jars. Another set of jars were left open. Redi's experiment was a clear success. In his "observations on the generation of insects," he recorded that he "began to believe that

4,200	4,100	4,000	3,900	3,800

ARCHEAN EON

Early seas.
Oldest rocks dated by radioactivity of Moon rock.

Exuberant volcanism and meteoric cratering continues.

Beginning of Earth crust formation and presumed start of tectonic activity.

Earliest Earth rocks (zircons from Mount Narryer in present-day Australia and Acasta gneiss from present-day northwestern Canada) dated by radioactivity.

Oldest Mars meteor crater (estimated).

Origins of life in the form of bacterial cells.

Appearance of first kingdom: MONERA.

Anaerobic prokaryotes and therefore autopoiesis, metabolism and reproduction have evolved.

Greenstone Isua Belt (present-day Greenland), indicating possible biologically produced carbonate and reduced carbon.

all worms found in meat were derived from flies, and not putrefaction."[4] Redi, in other words, developed a theory of maggots. Having seen flies hovering around and entering the open (but not the closed) jars, he confirmed his suspicion that the sealed meats, despite their putrid stench, did not become "wormy." In phase two of the experiment he covered meat with a cloth that prevented flies from laying eggs. No vermin appeared. He concluded that, "Earth, after having brought forth the first plants and animals by order of the Supreme and Omnipotent Creator, has never since produced any kind of plants or animals, either perfect or imperfect; and everything which we know in past or present time she had produced, came solely…from seeds of the plants or animals themselves, which thus, through means of their own, preserve the species."[5]

Scientists are said to abandon theories as soon as they are contradicted by experiment. In fact, many do the reverse, ignoring awkward experimental evidence in an effort to save appearances. Nowhere, to paraphrase Mark Twain, is so much derived from so little as in the production of scientific theory from scientific fact. A century after Redi's experiment, the English naturalist and Roman Catholic priest John Tuberville Needham (1713-1781) collaborated with the early evolutionist Georges Louis Leclerc Buffon (1707-1788). Buffon, as keeper of the Jardin du Roi, the French royal botanical gardens, was author of the forty-four volume *Natural History*, read by many of the educated class, including Erasmus Darwin, Charles's grandfather. Together, Needham and Buffon performed an experiment designed to determine whether spontaneous generation applied to all of life. Boiling mutton broth, they carefully sealed it in a jar. Opening it a few days later, they saw copious growth, suggesting to them that spontaneous generation did apply to microbial life. Although absolutely misleading—because they failed to kill the boil-proof microbes—the experiment ironically confirmed Buffon's essentially modern notion that "organic molecules" could under certain conditions combine to produce microorganisms.

In 1768 Italian biologist Lazzaro Spallanzani (1729-1799) demonstrated that his illustrious predecessors Buffon and Needham had neglected to boil the broth sufficiently. Still, Spallanzani's tests did not satisfy Ernst Haeckel who believed prolonged heat destroyed a "vital principle" in the air. Not until French chemist Louis Pasteur (1822-1895) exposed boiled meat extract to air by means of a flask, whose long neck was bent down and then up, were vitalists defeated. Air, but not bacteria, yeasts, or any other sort of life, could rise against gravity to enter the zigzag passageway to the life-supporting broth. As soon as the glass was broken, and microscopic life could enter, growth on the broth began. No other explanation held: life came only from previous life that was begotten by still earlier life. And yet, the work of Pasteur, proving that life comes only from previous life, strongly suggested that only God could have created life in the Beginning.

Origins of Life

In 1871 Darwin mused that one "could conceive in some warm little pond, with all sorts of ammonia and phosphoric salts, light, heat, electricity, etc." a chemically formed "protein compound…ready to undergo still more complex changes."[6] To trace life

3,700	3,600	3,500	3,400	3,300
First appearance of banded iron formation (BIF), suggesting local sources of oxygen at sediment-water interfaces.	Barberton Mountain Land (present-day South Africa) and Pilbara Block (present-day Western Australia), containing fossil evidence for anoxygenic communities: microfossils, stromatolites, and chemical fossils.	Onverwacht Group and Warrawoona Group (present-day South Africa), containing abundant reduced carbon in shales, microfossils, and stromatolites, imply widespread occurrence of photosynthetic bacterial communities. Earliest direct evidence of tectonic activity: granite of the Kaapvaal Craton from present-day South Africa.	Development of thickest (and therefore oldest) portions of continents.	Trace amounts of oxygen gas (O_2) in atmosphere and sediments.

back to matter was a logical extension of the idea that all species had evolved from a common ancestor. If species could evolve, what was to stop matter itself from evolving into life?

A young Russian biochemist, Alexander Ivanovich Oparin (1894-1981), published a book in 1929 entitled *The Origin of Life*. Oparin focused attention on specific ways in which chemicals might self-organize toward life. He described droplets growing by absorbing carbon compounds in a primeval soup. Theorizing an early hydrogen-rich atmosphere with gases such as methane and ammonia, and a solar source of energy, Oparin postulated that his "coacervates," or "semiliquid colloidal gels" would become increasingly dependent on their "own specific internal physico-chemical structure." Eventually,

> The internal structure of the droplet determined its ability to absorb with greater or less speed and to incorporate into itself organic substances dissolved in the surrounding water. This resulted in an increase in the size of the droplet, i.e., they acquired the power to grow.... A peculiar selective process had thus come into play which finally resulted in the origin of colloidal systems with a highly developed physico-chemical organization, namely, the simplest primary organisms.[7]

Because Oparin inhabited a nation (the former Soviet Union) that had been officially atheistic since 1917, he could theorize on this new version of spontaneous generation without confronting established religion.

In 1929 the British physiologist J. B. S. Haldane published an article making the point that reactive oxygen would have destroyed organic compounds; developing life, therefore, must have arisen in an oxygen-free atmosphere.[8] The work of Haldane and Oparin was an inspiration to "origins-of-life" experimenters from the United States, such as Stanley L. Miller, Sidney Fox, and Cyril Ponnamperuma. Nonetheless, Oparin was no more removed from his sociocultural milieu than were his predecessors; after World War II he declared Schrödinger's book *What is Life?* to be "ideologically dangerous" and he protested the new emphasis on genes, viruses, and nucleic acids, calling it "mechanistic." Yet Oparin, by imagining how life could have first evolved, revived the notion of spontaneous generation of life from nonlife.

In 1959 organic chemist Sidney Fox and his colleagues cooled water-free mixtures of amino acids to make "proteinoid microspheres." Resembling cocci bacteria, these microspheres would, under pressure, occasionally divide. Leslie Orgel of the Salk Institute in California discovered a DNA-like molecule (fifty nucleotides long) that formed spontaneously from simpler carbon compounds and lead salts. Five years later, ATP—the compound that is universally used by life to store energy—was produced by Carl Sagan, Ruth Mariner, and Cyril Ponnamperuma in a phosphorus-containing mixture of gases thought similar to Earth's early atmosphere. "It is, perhaps, ironic," writes University of Maryland chemist Ponnamperuma, "that we tell beginning students…about Pasteur's experiments as the triumph of reason over mysticism yet we are coming back to spontaneous generation, albeit in a more refined and scientific sense, namely, to chemical evolution."[9]

3,200	3,100	3,000	2,900	2,800

Continental tectonic activity—many small plates.

Formation of Fig Tree Group (present-day South Africa) of rocks that contain microfossils of reproducing cells.

Oldest evidence for life in present-day North America: Steep Rock, Ontario.

Widespread stromatolite reefs preserved at Steep Rock and Pongola Belt (present-day South Africa).

Diversification of bacteria—probably all major metabolic modes evolved by now (e.g., chemoautotrophy such as H_2, H_2S, NH_3, and CH_4 oxidation; oxygenic photosynthesis; reduction of iron and manganese oxides to metals.)

Gold deposited in paleoriver in present-day Witwatersrand, South Africa, indicating bacterial-mediated gold precipitation in ancient estuaries.

Large continents formed from raised portion of the plates known as the "preCambrian shields."

The "abiotic production" of ATP was actually a continuation of work begun by Stanley L. Miller, a graduate student of Nobel laureate Harold Urey, at the University of Chicago in 1953. Miniaturizing what he thought was Earth's earliest ambience, Miller filled flasks with gases (imitation atmosphere) over the surface of sterilized water (imitation ocean). For a week he bombarded his glassware microcosm with a lightning-like electrical discharge. The result was the evolutionary biologist's version of the just-twitched limbs of Mary Shelley's Frankenstein monster. Alanine and glycine, two chemicals essential to living proteins, as well as many other compounds, had spontaneously appeared in the flasks. In the laboratory, cooking from scratch, scientists had thus repeated the prebiotic origin not quite of life but of the nutrients needed for self-maintenance—a sort of primeval food.

Miller's laboratory mock-up of the early planetary atmosphere contained hydrogen gases like those left over from the gravitational accretion of the sun: hydrogen (H_2), water vapor (H_2O), ammonia (NH_3), and methane (CH_4). The experiments showed in startling fashion that the chemicals of life do self-organize without conscious direction. Given favorable conditions—Miller's model of the early atmosphere was only a rough guess—organic compounds spontaneously form from simpler precursors. The undeniable conclusion was that at least the matter of life spontaneously generates.

Miller's experiments were repeated and modified by many enthusiastic chemists. Some used alternative sources of energy, such as ultraviolet radiation and heat. Akiva Bar-Nun, for example, generated "sonic booms" in the laboratory; he showed that even energetic sound waves make protein components from atmospheric gases. Adenine, cytosine, guanine, thymine, and uracil—the five nucleic acid bases that strung together make DNA or RNA molecules—all have been synthesized in "prebiotic chemistry" experiments.

Of the six kinds of atoms crucial to life on Earth—carbon, nitrogen, hydrogen, oxygen, sulfur, and phosphorus—all have been detected in space. [PLATE 16] Hydrogen, the most common element in DNA, RNA, proteins, fats, and other compounds created by life, is also the most common in the universe. Ammonia (NH_3) was discovered in interstellar space in 1968. H_3C_2N, cyanoacetylene, was detected in 1970. Alcohol (CH_3CH_2OH) abounds in the constellation Orion. Other compounds found both in space and in living things include water, acetylene, formaldehyde, cyanide, methanol (wood alcohol), and five-atom formic acid, the clear fluid secreted by agitated ants. "God," declared Ponnamperuma, "is an organic chemist."

Ponnamperuma thinks our planet may have been "knee-deep" in polyaminomaleonitryl, an organic compound whose combinations could have inaugurated the later world of cells. Polyaminomaleonitryl is a polymer, a large molecule made of repeated links of HCN, hydrogen cyanide. HCN, a simple three-atom compound, has been detected on Titan, Saturn's sixth (and biggest) moon. A precursor to other biochemicals, including adenine and guanine in the nucleotide bases of RNA and DNA, HCN may be a key ingredient in the cosmic recipe for life. Ponnamperuma casts it as the "God molecule." The polymer comes in a range of colors, including the reds and browns characteristic of

PROTEROZOIC EON

Stromatolites abundant and cosmopolitan on ancient continents in parts of present-day Africa, North and South America, Australia, and Asia.

End of major crust-forming period.

Geologically modern processes begin: Oxygen gas (O_2) begins to seasonally accumulate; banded iron formations (BIFs) conspicuous and abundant; extensive huge lakes or oceans; carbonate platforms, indicating biogenic reef-like structures made by bacterial communities in marine settings.

First super continent (prePangea).

Beginning of worldwide age of BIFs: 90% of Earth's current mineable iron deposits in present-day southern Africa, Brazil, Central America, western Ontario, northern Michigan, and Minnesota formed between 2,400 and 1,800 mya.

Continued expansion of carbonate reef-like platforms and BIFs.

16
Triton, a photographic montage composed from a dozen different images of this moon of Neptune. Many of the outer planets and their moons possess water ice and carbon compounds like methane and carbon monoxide that seem more conducive to the formation of life than planetary and lunar environments in the inner solar system

2,200	2,100	2,000	1,900	1,800

Widespread occurrence of prokaryotic plankton (bacterioplankton) in world's oceans.

Increasing UV (ultra-violet ray)- absorbing ozone shield (O_3 derived from O_2 accumulating in atmosphere.)

Oldest abundant fossil bacteria: *Gunflintia, Huronospora, Leptoteichus golubicii*, etc.

Free O_2 abundant in atmosphere, indicating dominance of aerobic organisms.

Mitochondria, ancestors to most eukaryotes, acquired by symbiosis as purple eubacteria.

Gunflint Iron Formation (present-day Ontario, Canada) and equivalent fossil biotas in present-day China, Australia, and California containing complex filamentous microfossils and remains of structured communities.

First appearance of Grypania, identified as the earliest Protoctista (maybe reinterpretable as discarded cyanobacterial sheaths).

Replacement of banded iron formations by red beds (oxidized iron sediments), indicating worldwide transition to an atmosphere rich in oxygen.

"tholins," organic goop formed in the laboratory of astronomer Carl Sagan at Cornell University under conditions simulating those thought to exist on the similarly colored clouds of Jupiter.

Meanwhile, British chemist Graham Cairns-Smith has proposed that clays would have shielded fragile cell precursors from the solar rays which, though capable of assembling organic compounds, could also irradiate them into oblivion.[10] Crystalline clays, others have suggested, could have aggregated on nonliving bubbles produced by wind, rain, volcanoes, and waves. Even today, attracting particles along their surfaces while undergoing changes in temperature and pressure, bubbles serve as a meeting place where ambient carbon, nitrogen, hydrogen, and other elements form more complex compounds. When bubbles burst they leave chemical residues in their wake.

Whatever the precise route taken in the origin of life, Freeman Dyson proposes that it probably came by way of a kind of molecular "symbiosis" (although the word is not quite right, since neither partner was itself alive) between RNA—a "supermolecule" probably crucial, as we shall see, to life's origin—and more haphazardly growing "protein creatures."[11] Despite much conjecture and intriguing research, it must be remembered that no life has yet been synthesized in the laboratory. The gap between chemical evolution (the appearance of carbon compounds in "environmental" mixtures) and true cells (self-bounded, self-maintaining, and ultimately reproducing matter) remains unbridged. Nonetheless, the laboratory-explored shenanigans of RNA—as we shall see in a moment—are shrinking this bridge daily.

"Stumbling Forward"

Humanity's understanding today of the origins of life is probably no better than was our understanding fifty thousand years ago of the origins of fire. We can maintain and play with it, but we can't yet start it. The assumption that the origin of life may be repeated by investigators in the laboratory is a shocking example of the audacity of scientists—and yet it may prove correct.

Scientific investigation does reveal gradations between certain chemical systems and the animated material all of us recognize as life. Schrödinger's crystal analogy has given way to an idea of life as a chemical system requiring material and energy to persist far from thermodynamic equilibrium, i.e., a dissipative system. Dissipative systems that are not alive may nevertheless act in ways that are eerily lifelike. One such dissipative system develops in the Belousov-Zhabotinsky reaction. It involves the oxidation of malonic acid by bromate in a sulfuric acid solution containing cerium, iron, or manganese atoms. [PLATES 17 AND 18] Under certain conditions, concentric and rotating spiral waves will occur in an aesthetically captivating chemical reaction that may last for hours.

The regularity and duration of such reactions have led some scientists to compare them with life. Using energy from outside to increase their internal order, these chemical systems, some of which are brightly colored, "live" for a while beyond the limit of equilibrium chemistry. Erich Jantsch, an Austrian-American astrophysicist and philosopher, explains that,

> Whereas free energy and new reaction
> participants are imported, entropy and reaction

| 1,700 | 1,600 | 1,500 | 1,400 | 1,300 |

Appearance of second kingdom: PROTOCTISTA.

Earliest eukaryotes documented in fossil record as acritarchs, indicating cell evolution by symbiosis.

Origin of speciation inferred from molecular data on protoctists (primarily anaerobic mastigote protists).

Diversification of aerobic life.

Appearance of planktonic and benthic organisms possibly correlated to symbiotic acquisition of air-breathing mitochondria.

Protoctist evolution: origins of mitosis, meiotic sex, gender, and programmed death of individuals in eukaryotic microorganisms and their descendants.

Appearance of terrestrial cyanobacterial life (desert crust and soil microbial communities).

Diversification of seaweeds (algae, which are photosynthetic protoctists) of unknown taxa possibly correlated to symbiotic acquisition of photosynthetic plastids.

17 (above)
Proteus mirabilis, a bacterium. Phylum: Omnibacteria. Kingdom: Monera. Nested patterns of concentric terraces produced by *Proteus mirabilis.* The bacteria form these patterns by repeated cycles of growth and migration over the petri plate's agar surface. The living geometry of this photograph is reminiscent of the non-living dissipative structures of plate 18.

18 (below)
Autocatalytic chemical reactions of the same sort but using different substances as recursive structures are thought to give rise to life. This particular "chemical clock" is a dissipative structure in a Belousov-Zhabotinsky reaction. The increase in complexity over time is reminiscent of life. By reproducing, however, life has increased its complexity not for several minutes but for several thousand million years.

end products are exported—we find here the *metabolism* of a system in its simplest manifestation. With the help of this energy and matter exchange with the environment, the system maintains its inner non-equilibrium, and the non-equilibrium, in turn, maintains the exchange processes. One may think of the image of a person who stumbles, loses his equilibrium and can only avoid falling on his nose by continuing to stumble forward. A dissipative structure continuously renews itself and maintains a particular dynamic régime, a globally stable space-time structure. It seems to be interested solely in its own integrity and self-renewal.[12]

Dissipative structures, chemical systems that use streams of energy to increase their internal order, are, however, rare and short-lived. But if the increased internal order is that of life, then, given access of the system to a source of energy and the right kind of matter (nutrients), it maintains indefinitely. This is autopoiesis. Autopoiesis is what happens when a self-bounded chemical system—based not on small molecules of sulfuric and malonic acids but on long-molecule nucleic acids and proteins—reaches a critical point and never stops metabolizing.

The cell is the smallest autopoietic structure known today, the minimal unit capable of incessant self-organizing metabolism. The origin of the tiniest bacterial cell, the first autopoietic system, is obscure. Yet most agree that complex carbon compounds, exposed in some way to unceasing energy and environmental transformation, became oily droplets that eventually became membrane-bounded cells.

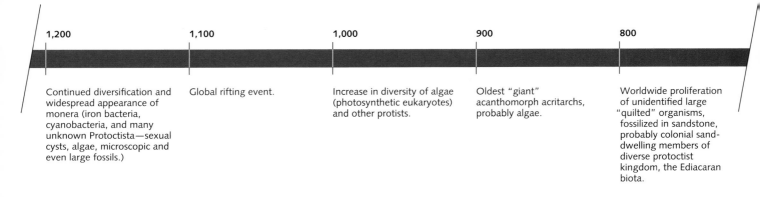

1,200	1,100	1,000	900	800

Continued diversification and widespread appearance of monera (iron bacteria, cyanobacteria, and many unknown Protoctista—sexual cysts, algae, microscopic and even large fossils.)

Global rifting event.

Increase in diversity of algae (photosynthetic eukaryotes) and other protists.

Oldest "giant" acanthomorph acritarchs, probably algae.

Worldwide proliferation of unidentified large "quilted" organisms, fossilized in sandstone, probably colonial sand-dwelling members of diverse protoctist kingdom, the Ediacaran biota.

Metabolism, the chemical measure, the specific earthly manifestation of autopoiesis, has been a property of life since it began. The first cells metabolized: they used energy (from light or from a small range of chemicals—never from heat or mechanical movement) and material (water and salts, carbon, nitrogen, and sulfur compounds) from outside to make, maintain, and remake themselves. Autopoiesis, the chemical basis for the impatience of living beings, is never optional. Absolutely required at all times for any life form in a watery milieu, once autopoiesis appeared in the tiniest bacterial ancestor, it was never completely lost.

You embody the processes of the early Earth in your living cells. The failure of the autopoietic system of cell maintenance is death. If autopoiesis of a cell ceases, the cell dies. A many-celled organism capable of replacing its cells survives as the autopoietic behavior of the larger organic being prevails. If too many component cells die, metabolism of the larger entity halts and death follows. Any cell or organism that continues to self-maintain will grow, and the imperative to reproduce will follow. Although not obvious to the naked eye, cell metabolism never stops. Chemical transformations such as nutrient uptake and energy conversion, and the fabrication of DNA, RNA, and proteins, occur continuously in all cells and all beings made of cells.

Life seems to have originated in whatever were the primordial ancestors of modern bacteria. Chemical systems that became biological systems, these first beings would have metabolized and incorporated energy, nutrients, water, and salts into their developing selves. The first cells formed. As in Jantsch's analogy of the person stumbling forward to avoid falling on his face, so membrane-bounded cells replicating RNA and producing other molecules stumbled onto DNA-based RNA and protein synthesis; that is, reproduction became a means to retain self-maintenance, to postpone return to thermodynamic equilibrium.

Bacteria reproduce in the time it takes to read this chapter. Elephant and whale reproduction can require a decade. But whatever the rate, reproduction requires DNA replication in cells. It requires RNA, protein, and membrane synthesis and the intrinsic locomotion of growth. Reproduction of larger beings—protoctists, fungi, animals, and plants—also involves growth and division of their component cells. Autopoietic multicellular beings are composed of cells which themselves are autopoietic. Animal and plant reproduction is a permutation of cell autopoiesis, just as cell autopoiesis is a permutation of nucleic acid and protein metabolism. Our instinctive desire to live is directly related to the autopoietic imperative to survive, itself related to the "yearning" of heat to dissipate.

Metabolic Windows

Because cells retain their organization in spite of—or because of—the helter skelter around them, they provide science with a window onto the past. It is a rather magical fact that, within the autopoietic, thermodynamic view, our bodies today should have virtually the same chemistry as that prevailing on Earth's surface three thousand million years ago. Remember that when life became autopoietic it postponed indefinitely the moment of total heat equalization and loss of order. Using the energy of food and sunlight, thermodynamic equilibrium has been thwarted.

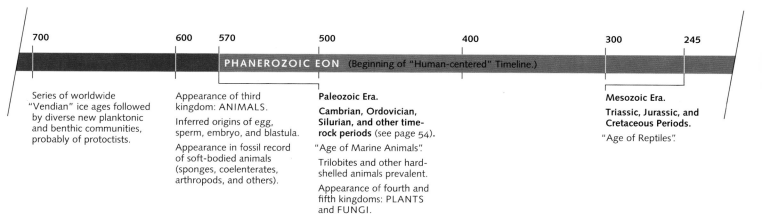

700   600   570   500   400   300   245

PHANEROZOIC EON (Beginning of "Human-centered" Timeline.)

Series of worldwide "Vendian" ice ages followed by diverse new planktonic and benthic communities, probably of protoctists.

Appearance of third kingdom: ANIMALS.

Inferred origins of egg, sperm, embryo, and blastula.

Appearance in fossil record of soft-bodied animals (sponges, coelenterates, arthropods, and others).

Paleozoic Era.

Cambrian, Ordovician, Silurian, and other time-rock periods (see page 54).

"Age of Marine Animals".

Trilobites and other hard-shelled animals prevalent.

Appearance of fourth and fifth kingdoms: PLANTS and FUNGI.

Mesozoic Era.

Triassic, Jurassic, and Cretaceous Periods.

"Age of Reptiles".

Death is illusory in quite a real sense. As sheer persistence of biochemistry, "we" have never died during the passage of three thousand million years. Mountains and seas and even supercontinents have come and gone, but we have persisted.

We have, of course, had to "up the stakes" at various junctures to stay alive. This continuous "upping of the stakes," which, on the personal level, links desire to death, is on the species level described as evolution. Beings always require food and energy to stay the same, and often they have to evolve, to change into new forms, simply to self-maintain. The feline lineage, the flowering plant lineage, the nautiloid-squid and the rest of the cephalopod lineage have changed and persisted through the sexual reproduction and death of their members.

Evolution, no less than the nucleic acid replication of autopoiesis and reproduction, is a "stumbling forward" to stave off the threat of thermodynamic dissolution. Most atoms in our bodies are hydrogen — the element which as a gas was, according to astronomic models, blasted beyond the confines of the inner solar system when the sun turned on. Nonetheless, these atoms which should be long gone have defied time and space by becoming bound up with (as) life. Today hydrogen-rich gases such as ammonia exist not only in the atmospheres of the giant outer planets but in the inner solar system where life has preserved them in its self-similar structure ever since it began maintaining and reproducing.

Indeed, the original dissipative chemistry, the protein and nucleic acid chemical clocks that arose prior to life may even have been preserved. One of the most beautiful aspects of living things is that they bear within their very form the presence of the past. We resemble our parents and other people who lived ten thousand years ago. This preservation of the past in the present is fortunate for scientists. Each body is the charitable gift of a biochemical museum, and each bacterial cell an unplanned time capsule.

Far from lost in what Shakespeare called the "dim backward and abysm" of time, life's origins are an open secret awaiting deciphering by sufficiently talented chemists. If life is an autopoietic, far-from-equilibrium phenomenon, living cells should still contain significant fragments of preliving systems. Vestiges of life's origin may still exist, a stuttering genesis for scientists patient enough to listen. Life may even contain the original dissipative structures and chemical fossils in the form of metabolic pathways. Ultimately far more valuable than micro-fossils, or the modern alchemical experiments of energizing chemicals in laboratory glassware, are organisms: easy to overlook in their obviousness — uncannily present — they are metabolic windows onto life's origin.

The RNA Supermolecule

The minimal free-living autopoietic entity today is probably a tiny, spherical, oxygen-shunning bacterium which requires energy and food to keep going its 3,000 genes and proteins. Or maybe it is a kind of mycoplasm, a being so small that until recently it was known only as a cousin to a growing speck that caused disease symptoms in brains of sheep. Even in these, the bonded atoms of carbon, hydrogen, nitrogen, and oxygen interact recursively in a metabolic system.

The genes that are DNA require active RNA to work. DNA and RNA together make the proteins that

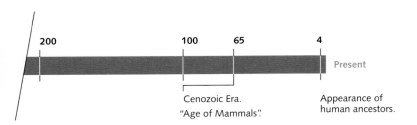

19 (next page)
The backbone of the feather-like structure is composed of protein-coated DNA actively being transcribed into messenger RNA molecules, which form the outward growing strands in this photograph taken with an electron microscope. This activity of the genes—DNA molecules using messenger RNA to make the thousands of proteins of all cells from the pseudopods of amebas to the feathers of parakeets—is the *sine qua non* of life as we know it on this planet. DNA copies itself to make genes but it is also transcribed into messenger RNA that puts together the rest of the organism.

form cell structures, and they also make the very enzymes that slice and splice the genes. The so-called genetic code actually refers to the correspondence between the linear order of DNA's components and that of the amino acids in a vast array of different proteins.

With the help of RNA, the nucleotides of DNA line up amino acids of protein. Our blood, internal organs, finger nails, skin, and hair are all made of proteins. [PLATE 19] The reason that nutritionists advise us to eat the "eight essential amino acids" is because the human body cannot renew itself without taking in these protein components from food and drink. The human body cannot synthesize these particular amino acids at all—even their simpler components.

In contrast to our human need to mine the environment for essential amino acids, no living being on Earth needs to stalk the environment in quest of the deoxyribose sugar essential for its DNA. Rather, deoxyribose is synthesized in cells by diverting an oxygen atom from ribose. It is ribose, the five-carbon sugar of RNA, that is often taken up from the outside as food. That all cells, given ribose sugar, can make deoxyribose from it suggests that ribose came first. RNA, with ribose, evolved before DNA. DNA sugar metabolism evolved by subtraction of oxygen from RNA sugars. The earliest cells may have been RNA beings which only later evolved DNA systems. Comparing RNA and DNA metabolism is an example of peering into cellular windows for clues to life's most ancient origins.

Other evidence questions the pretensions of DNA, the "master molecule," to life's biochemical throne. RNA, more versatile than DNA, is a better choice for the replicative tool of life's earliest autopoietic

system. Whereas double-stranded DNA uses deoxyribose sugar for its chain, single-stranded RNA uses ribose, the source material of deoxyribose. Unlike DNA, which must use RNA to code for proteins, RNA by itself can direct both its own replication and the making of proteins. In ancient times, RNA probably did all that DNA does today inside of cells, and more. In all cells, when the two helically coiled strands of DNA open to expose a section of the nucleotide sequence, that portion of the DNA is "copied" onto messenger RNA. Taking its message to two other kinds of RNA (transfer RNA and ribosomal RNA—named for the ribosomes, the "factories" in the cell where proteins are made), messenger RNA's information is "translated" into the amino acid units that assemble into working proteins. RNA can make proteins, in principle, without any DNA.

Following the lead of Sol Spiegelman at the University of Illinois in the late 1960s, German Nobel physicist Manfred Eigen (together with coworkers at the Göttingen Institute) found a way to induce test tube RNA molecules to replicate by themselves. Eigen showed that nucleotide units of RNA lined up and formed functional RNA. Most impressively, some of the test tube RNA even mutated into a different RNA that replicated more quickly than the original. The Eigen experiment did not, of course, reveal the spontaneous generation of life; RNA molecules by themselves are not cells. RNA in test tubes would have remained completely lifeless had scientists not extracted proteins from live cells and added them to test tubes containing RNA.

Eigen's RNA molecules are much like viruses. Certainly not alive, they show a power on the border produce all their cell parts. Others incorporate carbon dioxide and hydrogen from the air into body protein, converting their waste into methane gas. Still others turn sulfate to sulfide, or incorporate inert nitrogen into their bodies. Only citizens of the bacterial kingdom are so metabolically gifted. When an animal (like the termite who produces methane) or a plant (like the starved bean who begins to supply itself with nitrogen from its roots) is discovered with such metabolic skills it is because they have co-opted the bacterial bodies to their expertise. Such borrowing also applies to biotechnology performed by humans in white lab coats. We humans do not "invent" patentable microbes through genetic recombination; rather, we have learned to exploit and manipulate bacteria's ancient propensity to trade genes.

The Gene Traders

Bacteria trade genes more frantically than a pit full of commodity traders on the floor of the Chicago Mercantile Exchange. The trading by bacteria of genetic information provides the basis for understanding new concepts of evolution.

Evolution is no linear family tree but change in the single, multidimensional being that has grown now to cover the entire surface of Earth. This planet-sized being, sensitive from the beginning, has become more expansive and self-reflexive as, for the past three thousand million years, it has evolved away from thermodynamic equilibrium. Imagine that in a coffee house you brush up against a guy with green hair. In so doing, you acquire that part of his genetic code, along with perhaps a few more novel items. Not only can you now transmit the gene for green hair to your children, but you yourself leave the coffee shop with green hair. Bacteria indulge in this sort of casual quick-gene acquisition all the time. Bathing, they release their genes into the surrounding liquid. If the standard definition of

species, a group of organisms that interbreed only among themselves, is applied to bacteria, then all bacteria belong worldwide to a single species. The Archean Earth was a promiscuous place of prodigious growth and rapid gene transfer that led, by and by, to the genetic restrictions of the Proterozoic protists, the larger composite beings presented in our chapter 5.

Unlike all familiar sexually reproducing species, whose members have cells with nuclei in them that package their DNA, the DNA of bacteria is loose inside their bodies. Bacterial cells entirely lack nuclei; for this reason bacteria are prokaryotes composed of prokaryotic cells. "Prokaryote" literally means "before nuclei." Free of nuclei, and unfettered even by the red-staining, protein-coated chromosomes of all other forms of life, bacteria never reproduce by mitosis. Mitosis, the "dance of the chromosomes," is the kind of cell division by which the cells of plants, fungi, and animals always divide. This dance evolved in protists of the Proterozoic, the eon that followed the Archean. By contrast, a parent bacterium elongates its DNA dragged by growing membrane to which it is attached until the full-grown cell splits to form two offspring identical to it. Some bacteria reproduce by "buds," protrusions on the single parent that yield smaller offspring, all of which contain the parent's same genes.

Members of familiar species of plants and animals reproduce "vertically," as mother and father each donate an equal number of genes (on chromosomes) to form new offspring. Bacteria are under no such constraint. Rather, bacteria trade genes "horizontally," acquiring new genes from peers in their own generation.

Bacterial cells often have spare strands of DNA, that is, extra sets of genes. These genes may be traded in naked pieces called plasmids or as protein-coated pieces called viruses. In some bacteria a cell bridge forms between the one donating its genes and the one receiving them. [PLATE 21] This process of growing a cell bridge through which genes are sent, called conjugation, is distinct from mammalian sex. No bacterial cells fuse nor do "parents" make equal contributions to an offspring. Rather, one bacterium, the "donor," passes its genes in one direction to the "recipient," which does not reciprocate the favor. Still, this conjugation meets the minimal requirements of a biological sex act since the transfer of genes produces a new bacterium, a "genetic recombinant" being with genes from more than a single parent.

Bridge-forming bacterial conjugation is limited. Many types of bacteria that cannot conjugate indulge in viral or plasmid sex. Those that practice this more common form of sex require a difference between the bacterial "genders": a donor needs a recipient. Whether any given bacterial cell is donor or recipient is determined by a single "sex" gene. The sex gene may itself be transferred in the conjugation process. If this occurs, a "male" (donor) bacterium can become a "female" (recipient). "She" becomes a male donor like "himself." Any number from a very few to a very many genes may be transferred at a time, conferring on the recipient not just an ability to make cell bridges but other useful traits, such as an ability to manufacture vitamins or to resist a particular antibiotic.

When exposed to ultraviolet radiation, healthy bacteria explode with tiny viruses called prophages. Such viruses spread genes to surviving recipients. Because on the early Earth atmospheric ozone was not around to intercept the sun's ultraviolet rays, genetic exchange may have been even more prevalent than it is today. The early UV-bombarded Earth may have been the scene of a multimillion-year orgy of gene trading bacterial sex.

Bacterial recombination is a natural form of the "genetic" recombination exploited by biotechnologists. Manipulating a preexisting bacterial penchant, technicians force the colon

21
Three-way genetic exchange among bacteria. Unlike all other forms of life on Earth, bacteria transmit genetic information relatively freely such that taxonomically different "species" can trade genes. Bacterial sex, important to the evolution of cells with nuclei (eukaryotes) was probably rampant before bacteria themselves produced sufficient oxygen gas to create an ozone layer. The male in the upper left of the electron micrograph is sending genes through two tubes covered by bacteriophage viruses.

25
Alternating bands of hematite and magnetite, banded iron formations (BIF) are a source for much of the world's iron tools and machines. Such banded iron may have been formed by cyanobacteria growing in seasonal cycles of bloom and retrenchment. Ultimately so much oxygen was introduced by bacteria into Earth's atmosphere that iron precipitated not in bands but in rusty heaps called "red beds."

way to "red beds"—rust formations that formed all over the world. The rock record of oxidized minerals in Earth's crust testifies to the addition of oxygen to our planetary atmosphere in a 400-million-year stretch from 2,200 to 1,800 million years ago. Eventually no more minerals were left that had not already reacted with oxygen, so the excess gas with no place left to go began to accumulate in the air.

Quintessential Polluters, Quintessential Recyclers

In what passes for humility and respect for the ways of nature, modern humans worry about our pollution of Earth. Pollution is certainly distressing. But it is hardly unnatural. The pollution crisis effected by all-natural, blue-green bacteria was much worse than any we have seen lately. It destabilized the planetary environment. It made Earth inflammable, and to this day only the ancient surfeit of oxygen permits us to strike a match to make fire.

Human industry has increased the concentration of ozone-unfriendly chlorofluorocarbons in the atmosphere some one hundred times, up to about a billionth of a percent. This degree of change cannot even begin to compare with the effect upon the global environment wrought by the blue-greens. By growing, they increased atmospheric oxygen concentration from less than one part in 100,000,000,000 to one part in five (20%). And Earth's protective, ultraviolet-shielding layer of ozone (O_3, a three-oxygen molecule) was built up largely by "all-natural" pollution in the first place.

But if pollution is natural, so is recycling. Our fresh air is one-fifth oxygen. Today the ozone layer protects animals such as ourselves from ultraviolet skin cancer, cataracts, and compromised immune systems. One of the greatest turnarounds in evolution was the transformation of a once-fatal form of air pollution—oxygen—into a coveted resource.

Far from destroying the planet, oxygen energized it. In far-from-equilibrium systems waste products necessarily accumulate. But what may be garbage to one is dinner or building materials for another.

Bacteria, the greatest metabolic innovators, are not only the greatest polluters but the greatest cleaner-uppers. Our own chemical ability to use oxygen for energy derives from bacteria. Natural pollution recycling by bacteria extends to a host of other substances. Green and purple sulfur bacteria, starting with sulfide, produce sulfur globules and sulfate (both are more oxidized forms of sulfur), which suspend or dissolve in sea water. This sulfur is taken up and recycled by fermenting, sulfate reducing, or even other photosynthetic types of being.

Bacteria, in another one of their global megatricks, take nitrogen gas lost to the air and return it to all other living beings, where it is essential for the construction of proteins. Only a few types of bacteria own this miniaturized industry, as only a few are capable of breaking the strong triple bonds of molecular nitrogen and then sequestering the nitrogen atoms into organic molecules without oxygen sneaking in somewhere along the way. Bacteria thus "fix" gaseous nitrogen—by far the most abundant gas in the atmosphere—into organic compounds for all the living beings on Earth. Nitrogen-fixing structures, called heterocysts, were left 2,200 million years ago in the fossil record. These are large cells in chains mainly made of smaller ones. Cyanobacteria with heterocysts can fix N_2 gas and make it available as food. [PLATE 26]

The creative recycling metabolism of bacteria, combined with the imperative of autopoiesis, insures the biospheric flow of nitrogen, sulfur, carbon, and other compounds. Once nitrogen, for example, is fixed into protein and nucleic acid inside bacterial heterocysts, and once these proteins make their way through the food chain (degraded to amino acids and variously rebuilt along the way, with some leakage into the atmosphere as waste),

26
Fischerella, cyanobacterium. Phylum: Cyanobacteria. Kingdom: Monera. An example of bacterial metabolic "superiority," this cyanobacterium fixes atmospheric nitrogen in its heterocysts, making protein. The biochemical and metabolic repertoire of bacteria make them crucial to biological functioning on a global scale.

bacteria are summoned once again to do what only they can do: fix nitrogen back into organic molecules. The organically bound nitrogen in proteins and amino acids takes many routes. Some is degraded to ammonia (NH_3) by a diversity of bacteria. Ammonia is oxidized to nitrite (NO_2) or nitrate (NO_3) by still other bacterial specialists. Nitrite and nitrate, in turn, fertilize the water letting cyanobacteria and others grow. Nitrite and nitrate may be "breathed" by some bacteria which vent nitrous oxide ("laughing gas") and nitrogen (N_2) into the air. Nitrogen gas in the atmosphere must then be fixed again. The complex cycle never ceases. Although no bacteria yet degrade the refractory carbon-hydrogen compounds of most plastics, eventually some will evolve and, not limited by food supply, they will spread like wildfire from landfill to landfill through the biosphere.

Living Carpets and Growing Stones

Like magic carpets in certain remote corners of Earth "microbial mats"—huge numbers of interliving bacteria—have the power to take scientists back in time. [PLATE 27] Slick and slimy, often stinking of sulfur, microbial mats preserve the primeval scene of the early Earth before oxygen surfeited the air. The moist, multicolored mats found just inland from the sea feel cool beneath bare feet. Save for a sheet of algae, sand fly egg, occasional gull overhead, or paleobiologist footprint, nonbacterial traces are rare in microbial mats.

Modern microbial mats and scums are found everywhere in the world, but only in a few locations are they not obscured by larger forms of life. At Laguna Figueroa and Guerrero Negro in Mexico's Baja California, fronting the coastal town of Beaufort in the state of North Carolina, at Plum Island and Sippewissett in Massachusetts, and along the edge of the great Salt Lake in Utah and the sprawling Ebro river in Spain, mats are conspicuous. Visible in places too hot, cold, windy, or salty to

support larger life, microbial mats represent what life on a bacterial planet might have looked like several thousand million years ago.

Merging metabolic talents, a variety of bacterial forms organized into layers thrive in the mats that they themselves produce. The sun-loving cyanobacteria dwell in the upper layers, subtly and continuously transforming carbon, nitrogen, sulfur, and phosphorus—supplying these to their dependents below. Most are multicellular gliding filaments. Some are single-celled. Together, shaped as threads, spheres, or branches, the cyano colonies may form green jelly balls or scums in the water. Among the photosynthesizers, the most light-tolerant, desiccation-resistant forms inhabit the upper sheets of microbial mats; the sulfide-using and dim light-seeking members cohabit the lower layers. The purple sulfur bacteria navigate the middle realm, balancing a need for sulfide from the nether regions with a requirement for sunlight from the higher ones.

Lung-like, not only do the gases in the bacterial community move up and down daily, but so do the constituent guilds. Yehuda Cohen and his colleagues at the marine station at the Gulf of Eilat in Israel have determined that when the sun sets the purple layer of sulfide bacteria moves up a fraction of a centimeter. Deprived of sunlight, the blue-greens above no longer photosynthesize. When again at daylight the blue-greens begin bathing the bacteria below with their oxygen waste, that purple layer retreats.

Like a troop of sea turtles on the march—their shells jutting out of the shallow ocean in Shark's Bay, Australia—are the domed, layered rocks called stromatolites. [PLATE 28] The American geologist Charles Walcott called the fossil remains of very similar but very ancient rocks in Albany, New York, "cryptozoans" (from Greek for "hidden animals"). The locals in Saratoga Springs and around Skidmore College today call these formations "cauliflower

27
What appears as sludge in this beach-side lagoon is actually tightly packed bacterial communities. The upper layers grow photosynthetically, providing excess which feeds other bacteria in metabolically sophisticated ecosystems, once dominant on Earth, but now confined to areas too harsh, hot and salty for larger life forms. The microbial mat in this picture comes from Laguna Figueroa in Baja California Norte, Mexico.

28
Living bacterial stones called stromatolites from Shark Bay, Australia. Over time some types of microbial mats are thought to give rise to these strange domes. The photograph displays live stromatolites, replete with growing bacteria. Such structures are known from sea-side locations around the world. They occur in both fossil and live form. Bacterial land scapes were common before the evolution of animals, fungi or plants.

A petrified fossil stromatolite from Warrawoona,
South Africa (top) is compared with a cut section of
a living microbial mat,
Matanzas, Cuba (bottom).

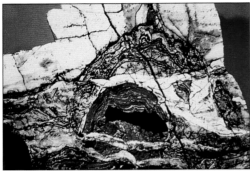

limestone." Although Walcott had an inkling that the round limestone "heads" were produced by life, only recently has it become clear that "cryptozoans" are stromatolites made by vast hordes of bacteria. They are, in essence, fossilized microbial mats that took the form of domes rather than the columns, reefs, and pancakes typical of the more spectacular ancient stromatolites. [PLATE 29]

The microbial communities, led by blue-greens, trapped, precipitated, and bound up calcium carbonate and grains of volcanic glass before they died. In Australia, where stromatolites are forming today by the work of live bacterial communities, the trapping and binding can be studied directly. Stromatolites

(which sometimes make use of environmental silica or even iron rather than only carbonate for building material) grow a layer at a time, as photosynthesizing bacteria glide past one another, slipping out of their polysaccharide sheaths, which are carbohydrate encasements similar in chemical composition to mucus. The sheaths are sticky and they bind sand. The live cyanobacteria, gliding toward the sun, leave behind their sheaths to be colonized by other shelter-seeking microbes. Trapping sediment and precipitating carbonate out of water, some of these complex mat communities solidify to make living fortresses against spume and wave. These fortresses thrive, as many kinds of photosynthetic bacteria support a plethora of camp followers. Spirilla, spirochetes, coccoids, and spore-formers in lively communities jostle for space, food, and position.

Some fossil stromatolites (such as those in the rocks of the Pongola Group in Africa, the Warrawoona Group in the Pilbara region of Western Australia, and the Swaziland in South Africa) harbor microscopic bacterial imprints. These silica stromatolites, made of black chert, are noteworthy for containing the microfossils that provide the best evidence for life in the Archean eon.

Thus, among their other accomplishments, bacteria created hard structures two thousand million years before the first animal evolved. Stromatolite hillocks would have been a common scene in the late Archean eon. Like miniature cathedrals, they were an early manifestation of life's ability to manage its excess. A landscape similar to that seen at Shark Bay has been in continuous existence somewhere on Earth since life began.

On a global scale the living tissues made by microbial mats—whether as living carpet or growing stone—may be as important to biospheric functioning as lung and liver are to us. Bacteria took over the world and still run it, using their decentralized planetary metabolism and capacity for worldwide intraspecies gene transfer.

So, what is life?

Life is bacterial and those organisms that are not bacteria have evolved from organisms that are. By the end of the Archean eon every desert was encrusted with microbial mats and temporary scums; every hot pool, sulfurous or ammoniacal, boasted hoards of colonists and pushy immigrants. Over salt grains and in rusty pools bacteria fabricated glues and precipitated magnetite. Clinging to the cold, barren rocks near the poles and sliming over the volcanic rubble in the tropical shallow seas, greening the Earth, photosynthesizers exuded their wares to hungry opportunists. The waste of a fermenter became the food of the acid-loving swimmer, while the fetid breath of a sulfate reducer provided a precious raw material to green chlorobia or red chromatia. Every available piece of real estate on this planet was occupied by enlightened producer, busy transformer, or arctic explorer. Naturally selected offspring survived, but only if lent a plasmid-borne gene from a community member. Gene exchanges were indispensable to those which would rid themselves of environmental toxins: a protein to be degraded, a poisonous manganese scum, or a threatening copper sheen to be oxidized or reduced. Replicating gene-carrying plasmids owned by the biosphere at large, when borrowed and returned by bacterial metabolic geniuses, alleviated most local environmental dangers, provided said plasmids could temporarily be incorporated into the cells of the threatened bacteria. The tiny bodies of the planetary patina spread to every reach, all microbes reproducing too rapidly for all offspring to survive in any finite universe. Undercover and unwitnessed, life back then was the prodigious progeny of bacteria. It still is. ■

5

Permanent Mergers

I have also seen a sort of animalcule that had the figure of the river eels: These were in very great plenty, and so small withal that I deemed 500 or 600 of them laid out end to end would not reach to the length of the full grown eel such as there are in vinegar. These had a very nimble motion, and bent their bodies serpentwise, and shot through the stuff as quick as a pike does through water.

—**Anton van Leeuwenhoek,** 1681

We cannot fathom the marvelous complexity of an organic being; but on the hypothesis here advanced this complexity is much increased. Each living creature must be looked at as a microcosm—a little universe, formed of a host of self-propagating organisms, inconceivably minute and as numerous as the stars in heaven.

—**Charles Darwin,** 1868

The greatest division is not even between plants and animals, but *within* the once-ignored microorganisms—the prokaryotic Monera and the eukaryotic Protoctista.

—**Stephen Jay Gould,** 1982

The appearance of these [protoctist] cells a billion-odd years ago was the second major event in planetary evolution and led directly, lineage by lineage, to our own complex selves, brain and all.

—**Lewis Thomas,** 1990

The Great Cell Divide

Some 2,000 million years ago, probably at many different sites on Earth, a new kind of cell evolved from bacterial interactions. The evolution of these complex new cells from integration of bacterial symbionts prepared the way for life in the new, Proterozoic eon. These new cells were ultimately the result of hunger, crowding, and thirst among teeming bacteria. These new cells were the first protoctists, and their coming brought the kinds of individuality and cell organization, the kind of sex, and even the kind of mortality (programmed death of the individual) familiar to us as animals. Bacteria merged. Curbing their viciousness and surrendering independence, they explored new ways to persist and reproduce.

Our kind of life, that of the nucleated cell, began long before animals. Amid cell gorgings and aborted invasions, merged beings that infected one another were reinvigorated by the incorporation of their permanent "disease." The first new kind of cell—the nucleated cell—evolved by acquisition, not of inherited characteristics but of inherited bacterial symbionts. These new kinds of cells making up the bodies of unicellular protists and multicellular protoctists would eventually lead to the final three kingdoms of life yet to evolve on Earth: animals, fungi, and plants. Our protoctistan ancestors were beings so exceedingly weird that, if informed in detail of their existence, even the credulous author of a medieval bestiary might pooh-pooh the tale as the impossible product of a febrile imagination.

Each and every organic being on Earth is made of one of only two kinds of cells. Our kind—and that of other animals, fungi, plants, and protists—possesses nuclei. The other kind, the bacterial cell, has no nucleus. In 1937 Edouard Chatton, a French marine biologist, named the latter cell type "procariotique"; the organisms that possess this cell type are prokaryotes (pronounced, "pro-CARRY-oats"). All the rest of us are eukaryotes ("you-CARRY-oats"), made of nucleated cells. So the presence of a membrane-bounded nucleus defines a cell as "eukaryotic." All eukaryotes come from protoctists; bacteria don't. The long DNA molecules, the genes of eukaryotes, are organized within the nucleus into at least two and as many as several thousand chromosomes (humans have forty-six). As we shall see, this sequestering of precious genetic material inside a special membrane and the firm binding of DNA into a particular sequence in a particular chromosome limited the genetic promiscuity that was and still is accepted practice in the bacterial realm. [PLATE 30, *following page*]

A giraffe is a eukaryotic organism, made of eukaryotic cells. So is a daisy. And an ameba. The differences in behavior, genetics, organization, metabolism, and especially structure between prokaryotes and eukaryotes are far more dramatic than any between plants and animals. Those differences mark the great cell divide. Prokaryotes and eukaryotes thus form the two "supergroups" of life on Earth.

All of one supergroup and a good portion of the other inhabit the microbial realm. Bacteria, the smaller protoctists, and yeasts and other small fungi are microbes. The eukaryotic cells of protoctists and fungi are bigger than the prokaryotic cells of bacteria; but, like any cell, they must be viewed with a microscope. The path of transition between the two supergroups is obscure. The evolution from prokaryotes to eukaryotes, from bacteria to protoctists, was a "symmetry break" that catapulted life to a greater level of complexity and gave it different potentials and risks. Not just by gradual mutation but suddenly through symbiotic alliance did the first eukaryotes form.

Five Kinds of Beings

The earliest eukaryotic cells, living on their own, were protoctists that evolved by permanent bacterial merging. Floating or free-swimming, some went on to become animals, fungi, and plants.

The protoctists are a wide-ranging group of obscure beings. Today an estimated 250,000 species include tiny amebas and diatoms and giant kelps and red seaweeds. Ultimately, this group gave rise to familiar plants and animals such as palm trees and clams. But even as recently as a thousand million years ago not a single animal, plant, or even fungus dwelled on Earth. Biospheric functions were handled entirely by bacteria and protoctists.

The ungainly name "protoctist" was introduced by an English naturalist with the equally unenviable name: Hogg. John Hogg (1800-1861) set forth his views in an article published in 1861, just before he died: "On the Distinctions of a Plant and an Animal, and on a Fourth Kingdom of Nature."[1] (His third was the "mineral kingdom.") Neither Hogg nor anyone else at that time was aware of prokaryotic and eukaryotic cells. But Hogg saw that many organisms were neither plant nor animal.

Unlike the term protozoa ("first animals"), with its unfortunate connotation that organisms ranging from foraminifera to slime nets were somehow animals, protoctist simply means "first beings." Protoctists are neither animals nor necessarily single-celled. But when they are single-celled—or otherwise tiny—they are called protists. Because all animals grow from multicelled embryos, there are, by definition, no single-celled animals. So-called single-celled animals are really the protists, the smaller protoctists. Hogg suggested "Regnum Primogenium" as the name for this primordial kingdom. Its founding members are now known to have originated prior to plants and

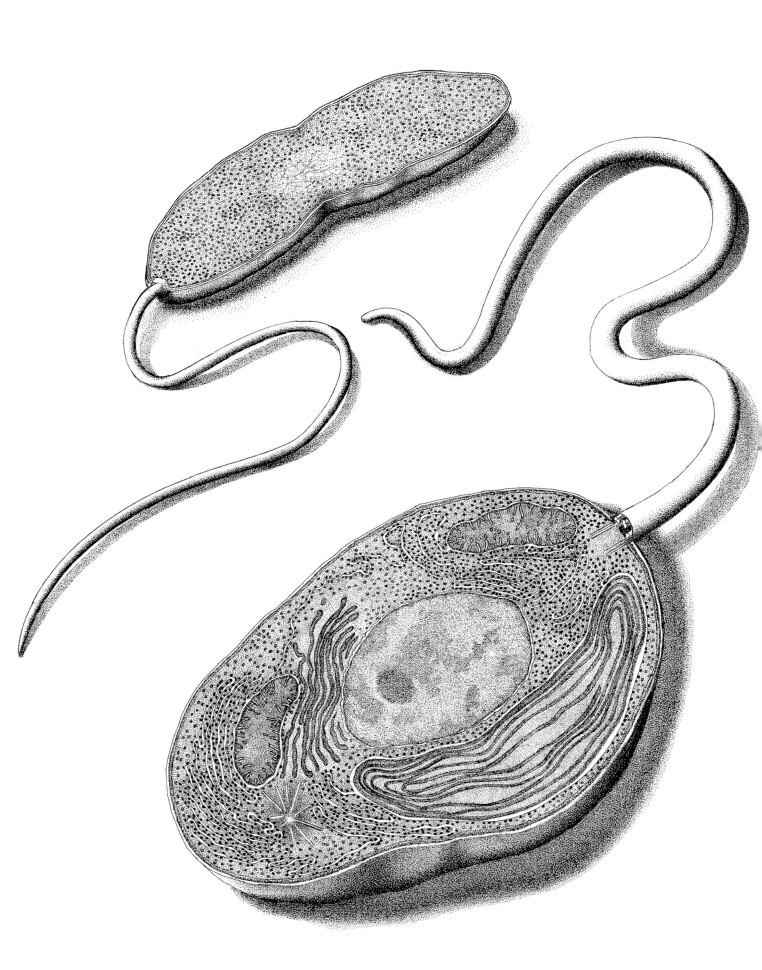

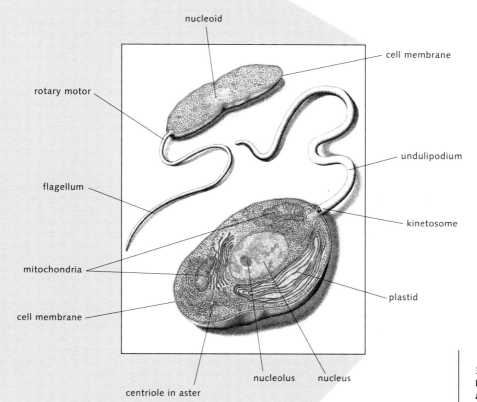

nucleoid

cell membrane

rotary motor

undulipodium

flagellum

kinetosome

mitochondria

cell membrane

plastid

centriole in aster

nucleolus

nucleus

30
Illustrated comparison of a prokaryote—bacterium— at top and a eukaryote— nucleated cell—at bottom. All living cells on Earth are either prokaryotes or eukaryotes. The non-bacterial kingdoms— Protoctista, Fungi, Plantae and Animalia—all consist of organisms whose cells are eukaryotic. Eukaryotes evolved symbiotically from eating, invading, infecting, and cohabitating bacteria.

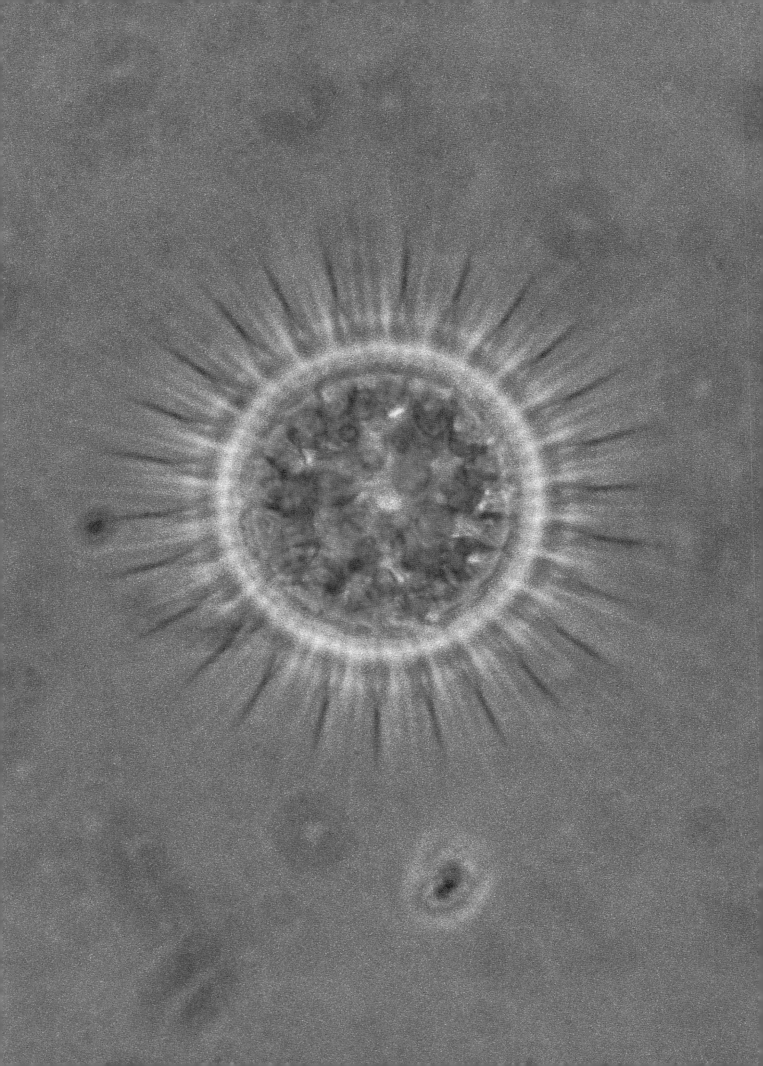

animals, and yet protoctists continue to thrive on Earth today. [PLATE 31]

In Germany Ernst Haeckel also argued for a new kingdom. "These interesting and important organic beings are the *primary creatures* or *Protista*."[2] The Monera—bacteria—were part of Haeckel's proposed Protista. Haeckel, recall, was not persuaded by Lazzaro Spallanzani's boiling of mutton broth to kill microbes. It seemed clear to him that primordial beings simpler than anything yet discovered must exist. An ardent believer both in evolution and in the spontaneous generation of matter, Haeckel sought "an entirely homogeneous and structureless substance, a living particle of albumin, capable of nourishment and reproduction."[3]

English biologist Thomas Henry Huxley (1825-1895) was taken with Haeckel's notion of a primordial protein globule. Examining ten-year-old mud samples dredged up from the seabed off the northwest coast of Ireland, Huxley discovered a mysterious white ooze. Were these Haeckel's postulated earliest Protista? Upon examination, the granular ooze was seen to consist of tiny calcareous plates. Excited, Huxley wrote to Haeckel that he had encountered the ancestral life form. Indeed in the flush of his discovery, Huxley honored his colleague by naming the "organisms" after Haeckel. Both men delivered the exciting news that *Bathybius haeckelii*, the great Urschleim (primordial goop), had finally been found.

Only later was it realized that *Bathybius haeckelii* was just marine sediment. The white slime that appeared whenever Huxley doused the ooze to preserve it was an alcohol precipitate of organic debris that included jellyfish stingers. Far from being our primordial parent, the Urschleim was not even alive. Nevertheless, Haeckel's concept focused scientific attention on beings that escaped the dichotomous plant/animal classification scheme.

Today the tendency to divide life into animal versus plant remains. Fungi, if they exist at all in the

31
Mesodinium rubrum, a protist. Phylum: Ciliophora. Kingdom: Protoctista. This fast-swimming but photosynthetic microscopic being dwells in brackish water near the Baltic Sea. It exemplifies organisms confounding older, three-kingdom classification systems since it is really neither plant nor animal. The reddish hue of the interior derives from degenerate symbiotic algae.

popular imagination, are a kind of gray plant. Smaller protists and bacteria—not quite life in the popular mind—are ignored or lumped together as "germs." Academia still departmentalizes life into botany, the study of plants, and zoology, the study of animals. Fungi, bacteria, and certain protoctists are often forced in this scheme to be plants under the jurisdiction of botanists. This quaint plant-animal split does not reflect evolution. The ancestors to plants and animals were neither; rather they were communities—bacteria that merged to form a new kind of cell.

The first essentially modern classification was invented by Herbert F. Copeland (1902-1968), a biology teacher at Sacramento City College in California. Copeland argued for four kingdoms: Monera (bacteria), plants, animals, and protoctists. He placed all fungi (molds, mushrooms, puffballs, etc., which he called "Inophyta") into a subdivision of Hogg's Protoctista. His book, *The Classification of the Lower Organisms*, published at Copeland's own expense by a so-called vanity press, was read by almost no one except Cornell University ecologist Robert H. Whittaker (1924-1980). Whittaker devised the most useful groupings of all when he removed fungi from his Protista and recognized them as a distinct "fifth kingdom."

From today's vantage, Whittaker's five-kingdom classification scheme best reflects evolutionary relationships. One of us (Lynn Margulis) has collaborated with zoologist Karlene Schwartz of the University of Massachusetts at Boston to sharpen the blurred boundaries of Whittaker's protists. The Kingdom Protoctista, which Whittaker limited to unicellular and the smallest multicellular beings, now includes larger organisms that are not plant, animal, fungi, or bacteria, such as seaweeds.

Twists in the Tree of Life

The story of how a human—a being made of nucleated cells—evolves from an ameboid being—a nucleated cell—is bizarre. But even this story has a preamble: the evolution of a cell with a nucleus. How did such a cell evolve?

The quick answer is by the merging of different kinds of bacteria. Protoctists evolved through symbiosis; twigs and limbs on the tree of life not only branched out but grew together and fused. Symbiosis refers to an ecological and physical relationship between two kinds of organisms that is far more intimate than most organism associations. In Africa, for example, plovers pluck and eat leeches from the open mouths of crocodiles without fear. Bird and beast in this instance are behavioral symbionts; crocodiles enjoy clean teeth in the company of well-fed plovers. Bacteria live in the spaces between our teeth and in our intestines, mites inhabit our eyelashes; all these tiny beings draw nutriment from our cells or our uneaten food, as cells are shed or as they excrete organic excess. Symbiosis, like marriage, means living together for better or worse; but whereas marriage is between two different people, symbiosis is between two or more different types of live beings.

Organisms form many kinds of symbioses, but the most awe-inspiring is the exceedingly close association known as endosymbiosis. This is a relationship in which one being—microbe or larger—lives not just near (nor even permanently on) another, but inside it. In endosymbiosis, organic beings merge. Endosymbiosis is like a long-lasting sexual encounter except that the participants are members of different species. Indeed, some endosymbiotic linkages have become permanent.

Bacteria, masters of symbiosis in general, are also the best endosymbionts for at least four reasons. First, because they have been entering into stable relationships with one another for several thousands of millions of years, they are good at forming permanent relationships. Second, their tiny bodies fluidly lose and acquire genes, making them amenable to rapid genetic change. Third, bacteria have only a limited expression of individuality; no circulating antibodies guard them—an "infection," far from being rejected as it might be in an animal with immune system, can thus become the basis for life-long association, a mutual evolution. Fourth, bacteria's vast chemical repertoire leads to a tendency for metabolic complementarity less often seen in associations between already highly individualized members of plant and animals species. Of course, given time, some plants and animals may come together as closely as some bacteria have.

Symbiosis produces new individuals. "We" could not synthesize B or K vitamins without bacteria in our gut. Cows and termites are not themselves without the swimming fermenters in their digestive systems—ciliates and bacteria that break down grass and wood. Some algae living inside translucent flatworms are such good providers that the worms have atrophied mouths; the close-mouthed green worms "sunbathe" rather than seek food, and the endosymbiotic algae even recycle the worm's uric acid waste into food.

Thousands of other strange partnerships exist. All of the estimated 20,000 lichens, for example, began as symbiotic associations of algae with fungi or of cyanobacteria with fungi. But the most important symbioses were those that led to the eukaryotic cell.

Today most protoctist cells and all plant, animal, and fungal cells contain mitochondria. The oxygen-respiration that keeps members of the youngest four kingdoms of life alive takes place inside these particular organelles. (Like organs within bodies, organelles are functioning structures within eukaryotic cells.) Mitochondrial organelles look like bacteria. They even grow and divide in two at their own pace within the larger cell. They are thought to come from bacteria—but after more than a thousand million years of association they cannot survive outside the confines of the cell.

The cells of plants and some protoctists such as algae also possess colorful bodies called plastids. All the photosynthesis undergone by algae and plants happens inside the DNA-containing plastid organelles. Plastids contain the same pigments and other biochemicals found in the spherical, oxygen-producing blue-green bacteria that thrive in the ocean. Coincidence? We don't think so. Indeed, DNA in the plastids of the cells of the red seaweed *Porphyridium* is closer in its nucleotide sequence to that of certain cyanobacteria than it is to DNA in the nucleus of the red seaweed itself.

Such genetic evidence links the cell organelles to their origin from free-living bacteria in a definitive (and now virtually undisputed) way. Genetic similarities that cross kingdoms is the biological equivalent of ancient "fingerprints," proving that photosynthetic organelles did not evolve gradually by a buildup of mutations in the DNA of plant and algal progenitors, but suddenly when symbiotic bacteria took up residence in larger cells. In a moment we will return to the question of how the bacteria that became mitochondria and plastids found their way to their current, cozy location inside the cell. But, to be chronologically correct, we must first explore what may be a still older, and deeper, symbiosis.

Squirmers

Nearly all biologists now accept that particular bacteria, after a period of chemical negotiation and gene transfer, began as symbionts and became the mitochondria and plastids of larger cells. Most biologists, however, reject or are ignorant of another idea. Nonetheless, circumstantial evidence suggests that a still older bacterial symbiosis preceded the acquisition of these organelles.

Before oxygen-users infected anaerobic, swimming protists to form alliances, and before blue-green bacteria were engulfed by these alliances, faster bacteria seem to have conjoined.

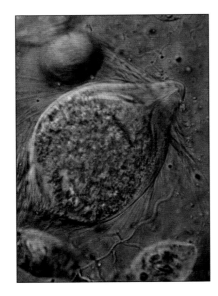

32
Trichonympha, a chimerical protist. Phylum: Zoomastigina. Kingdom: Protoctista. This being, as peculiar structurally as any to be found in a medieval bestiary, is composed of a large protoctist host and a swarm of both undulipodia (its organelles in the front) and symbiotically attached spirochete bacteria in the rear. It is itself symbiotic in the termites' hindgut, a microscopic zoo containing many different sorts of protists and bacteria that together aid in the digestion of wood.

Transforming from free-living bacteria to become parts of cells, wriggling spirochetes may have conferred their considerable powers of movement upon the outside, and then the inside, of victims that became ancestral cells. Today's spirochetes are proton-powered bacteria that ferment carbohydrates and whip about like possessed corkscrews. The most rapid swimmers of the entire moneran kingdom, they literally screw their way through mud, tissue, and slime.

Thriving in saliva, the crystalline-styles—digestive tissues—of oysters, the hindguts of termites, and a thousand other equally ingenious niches, invasive spirochetes are one of the most successful life forms on Earth. And they form alliances—often reversibly attaching to larger organisms and propelling them along. Some protist cells—such as *Mixotricha paradoxa* and *Trichonympha*—have gone so far as to evolve holdfast structures where free-living spirochetes are encouraged to reversibly "dock"—with their engines running. [PLATE 32] The spirochetes actively feed on the metabolic leftovers of the cells to which they attach. The symbiotic advantage is obvious: the squirming spirochetes move the cells that feed them. A *Mixotricha* or *Trichonympha* cell without spirochetes is like a boat lacking a motor or a teenager without a car. Consortia able to swim quickly have more opportunities than their sluggish predecessors to find food, escape predators, and meet mates. But external spirochetal attachments are not the whole story. [PLATE 33]

33
In this drawing artist Christie Lyons portrays spirochetes attaching and becoming the undulipodia of larger, new eukaryotic cells.

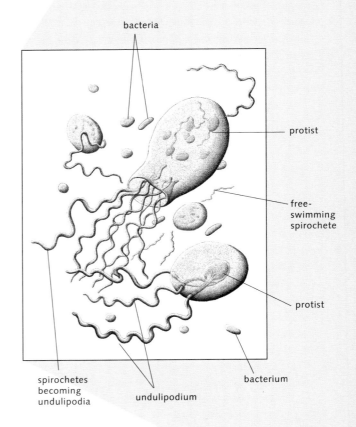

bacteria

protist

free-swimming spirochete

protist

spirochetes becoming undulipodia

undulipodium

bacterium

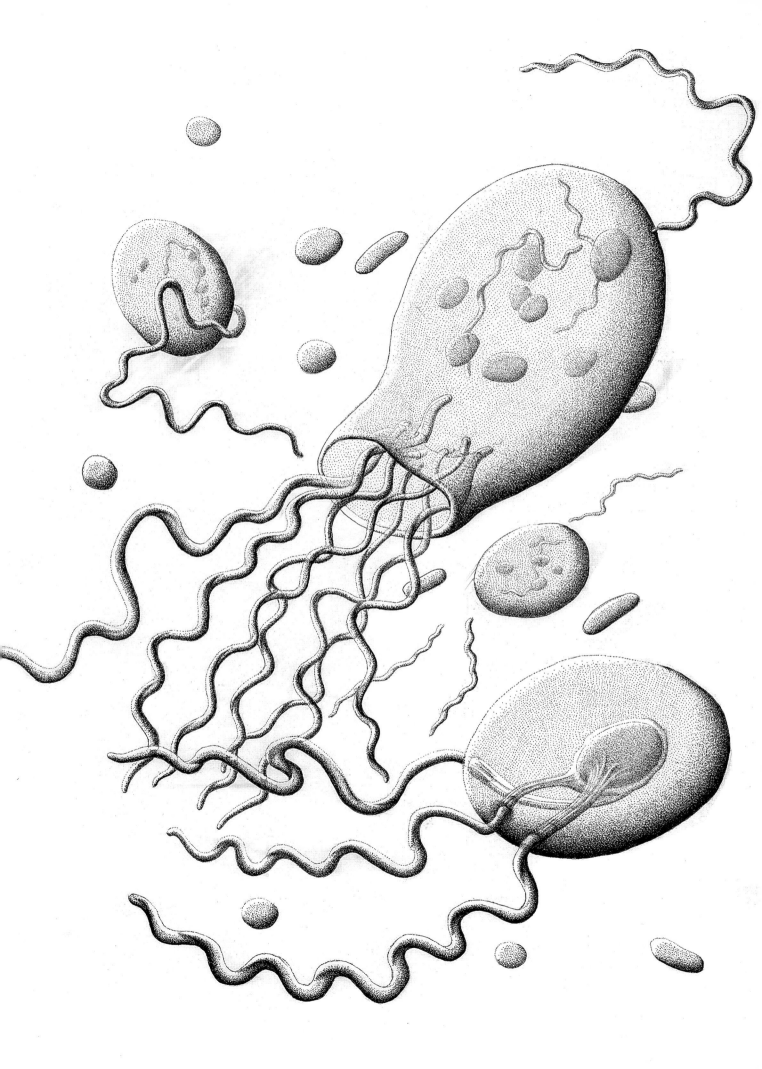

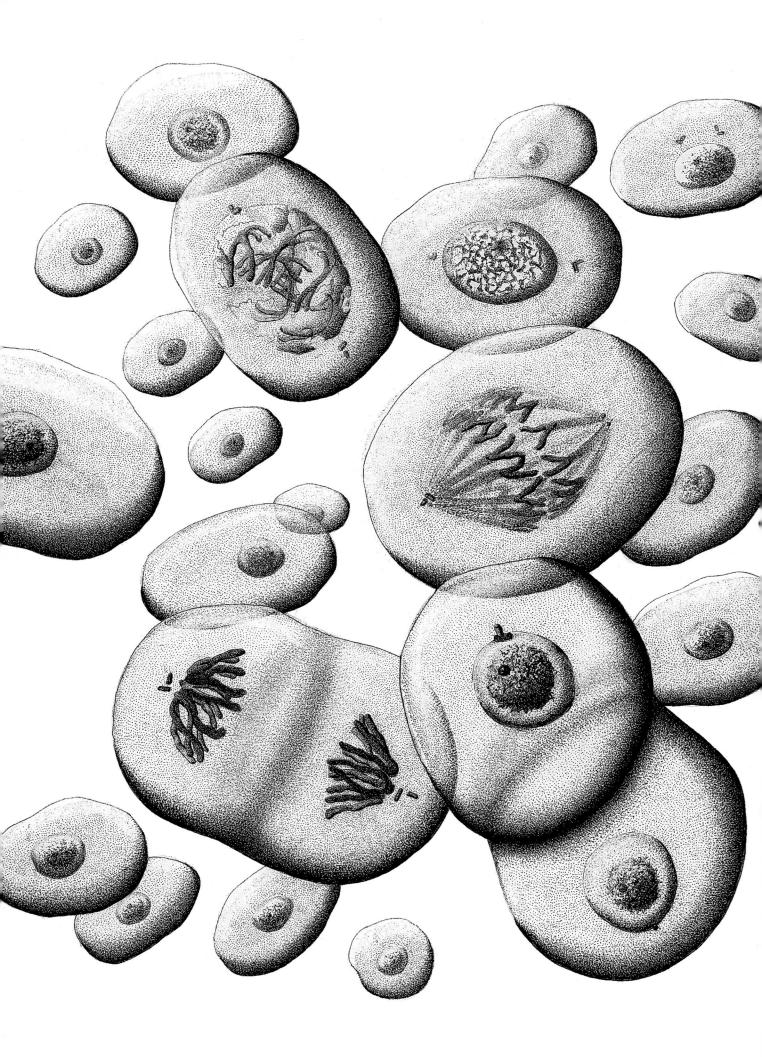

34
Different stages of mitosis, the usual method of chromosomal separation prior to cell division in reproducing eukaryotic cells. The greater internal movement of cells with nuclei as compared to bacteria may result from the 2,000 million-year-old remnants of rapidly squirming spirochetes.

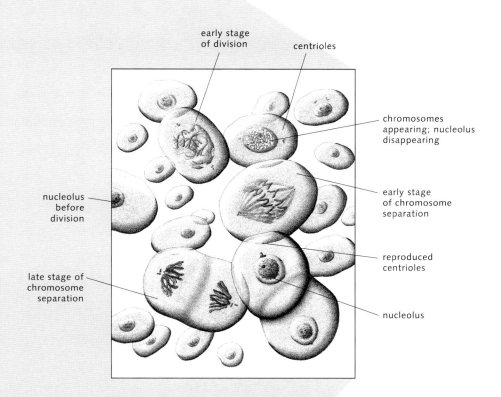

early stage
of division

centrioles

chromosomes
appearing; nucleolus
disappearing

nucleolus
before
division

early stage
of chromosome
separation

reproduced
centrioles

late stage of
chromosome
separation

nucleolus

Protoctist cells, huge compared to bacteria, display incessant internal movement. Bacterial cells, lacking any internal movement and real chromosomes, do not divide mitotically; they do not perform "the dance of the chromosomes." Mitosis, the chromosome style of cell reproduction, is widespread among protoctists, and universal in their animal, plant, and fungal descendants. Matching chromosomes line up and move to opposite poles in a kind of micro-ballet. At the mitotic poles in animal and many protist cells are centrioles, structures resembling rotary telephone dials that may be the remnants of spirochetes that long ago entered larger cells to feed. [PLATE 34]

The mitosis by which most eukaryotic cells divide ensures that chromosomes doubled in the parent are

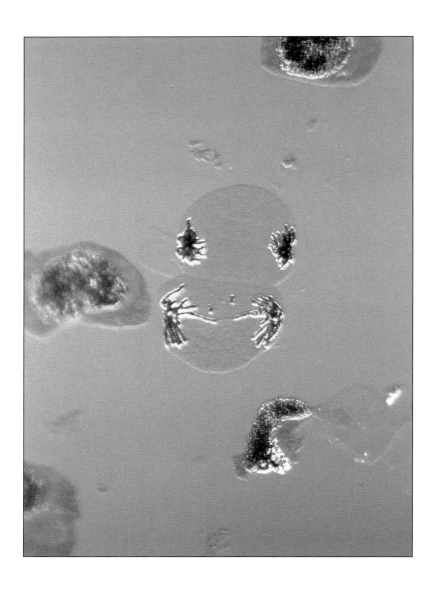

35

Haemanthus sp., African blood lily. Phylum: Angiospermophyta. Kingdom: Plantae. Telophase in the mitotic cells deep inside the flower. Mitosis is the kind of reproduction of cells by division that typifies the nucleated cells of all animals, plants, and fungi and most protoctists. Telophase is a late stage of mitosis. In mitosis the chromosomes double first and then divide and separate into these two masses that become the nuclei of the resulting offspring cells. In the mitotic cell division shown here the flower's chromosomes are stained red.

partitioned evenly into two offspring cells. Mitosis seems indispensable as a genetic filing and distribution system for the huge quantities of DNA that most eukaryotic cells contain. In each episode of mitosis a series of tiny protein tubes, microtubules (collectively called the mitotic spindle), appears. At the end of the process, when one cell has become two, this mitotic spindle disappears. The chromosomes attached to the tubules of the spindle line up along the plane of the cell's equator. These chromosomes, which doubled earlier, now separate as each half moves along the spindle to the opposite side of the cell. The chromosomes now at the poles uncoil as the cell proceeds to divide into two. The enigmatic mitotic spindle then fades back into the invisibility from which it emerged. [PLATE 35]

In many animals the centrioles (the telephone dial-like structures) move to the edges of the cell, where they become kinetosomes by growing shafts. In cross section, the shafts show a distinctive "9(2)+2" pattern: nine sets of two tubules arranged near the perimeter of the circular axis, with one set of two tubules at the center. The kinetosome and centriole have different names only to distinguish the shafted from the shaft-less phase of the same organelle. Multiple names for the same organelle are more accidents of history than nomenclatural necessities. Two names were bestowed because the distinct phases were noticed and named long before any relationship between them was recognized.

The universality of the kinetosome shafts — found in plants, animals, fungi, and protists — is strong evidence for an ancient origin. The 9(2)+2 symmetry is found, for example, in the cell extensions of the balance organ of our inner ears and in tails propelling the swimming protist *Euglena*. The 9(2)+2 arrangement can be seen in cross section in sperm cells of men. Because of their similarity, all 9(2)+2 shafts that grow from kinetosomes are best referred to by a common name. We call them undulipodia ("waving feet").

The cell reproduction of mitosis — remarkably similar in some protoctist and all plant, animal, and fungal cells — must have evolved in the oldest of these four kingdoms. The protoctists, from which emerged plants, animals and fungi, were the first beings with this motility apparatus necessary for the reproduction of the new cells with nuclei. But it seems doubtful that the smaller protoctists invented the undulipodium and internal cell movement. Motility, rather, could have been the gift of the oldest and original kingdom of life.

Tantalizing evidence implicates a bacterial origin of the centriole-kinetosome organelle. Both DNA and RNA have been reported in these intracellular structures. David Luck and John Hall at Rockefeller University in New York City photographed a peculiar, bacteria-like DNA in the two centriole-kinetosomes of the green alga *Chlamydomonas*. Joel Rosenbaum and his colleagues at Yale University, along with several other scientists working independently, have not, however, been able to confirm any centriole-kinetosomal DNA in this green alga.

Living cells are decorated by undulipodia that go by many different names. Undulipodia include all cilia and the "tails" of most sperm. The single-tailed swimmer in bull semen and the hundred-tailed sperm released by male fern plants are both examples of 9(2)+2 undulipodia. The immotile cilia remnants in the rod and cone cells of our retinas, the motile ones of the fallopian tube cells that push a woman's egg toward the womb, and those rejecting debris in our windpipes are further examples of undulipodia.

It may be that the spirochetes that symbiotically became undulipodia (involved both as cell tails and chromosome movers) have become so integrated with their partners that they have dissolved away to mere traces and genetic shadows of their former selves. Like an artist who seems effortless in the performance of a difficult routine, the former spirochete genes may be so deeply implicated in cell function that today they all but defy detection. Oxford University biologist David C. Smith likens such symbiotic remains to the smile of the Cheshire Cat, the fictional feline of Lewis Carrol in *Alice in Wonderland* who slowly fades away to become nothing but an enigmatic grin, floating in midair: "the organism progressively loses pieces of itself, slowly blending into the general background, its former existence betrayed by some relic."[4]

The remaining traces of the bequeathers of motility are fewer and hazier than those left by the cells which gave the green gift of photosynthesis and oxygen-bubbling respiration. Motility, in our view, was the first endosymbiotic acquisition of the nascent eukaryote; losing parts of themselves as they evolved, squirming spirochetes invaded and animated what were to become nucleated cells. Today, because of time, the evidence is thin. The photographs have all but faded and the pages have crumbled. Cell history must be reconstructed from the faintest clues.

A reason to think that spirochete symbiosis preceded the others is the recent discovery of many protists that have undulipodia but lack mitochondria. These air-shunning beings are poisoned by oxygen—suggesting they date from a time before ancestral protists had become symbiotic with the oxygen-using bacteria that evolved into mitochondria. Mitotic cell division in which chromosomes line up on the spindle is universal in animal, plant, and fungal cells. Only a few dark-dwelling, oxygen-shunning, swimming "amitochondriate" protists and their obscure relatives show important variation on the mitosis theme.

The absence of intermediates between bacteria and such seemingly aberrant protists tells us that evolution from bacteria to nucleated but still anaerobic swimmer probably did not occur by random mutation alone. The sudden evolution of cells with nuclei and 9(2)+2 swimming organelles is best explained by ancient motility symbiosis. When the close connection of undulipodia and mitotic apparatus is observed in live anaerobic cells, symbiosis becomes the most parsimonious of all scientific explanations. Indeed, by comparison, mutation explanations for the origin of undulipodia seem far-fetched.

Consider a very ancient ancestor of one of today's bacterial denizens of hot springs, *Thermoplasma* [see PLATE 35]. Imagine that ancestor under attack by spirochetes. Holding firm, its protective membrane resists penetration. The spirochetes attach on the outside, establishing association, as they feed on *Thermoplasma*'s waste. Eventually some gain entry and merge with the debilitated *Thermoplasma* to become its living oars.

Once inside, the spirochete symbionts extend their motility skills to the internal operations of their would-be victim. A sort of biochemical truce prevails, as both sorts of reproducing partner manage to coexist. The nucleus, acting today as a sort of central genetic government, might have evolved as membrane proliferated to keep the attacking spirochetes from eating out *Thermoplasma*'s DNA. The captive spirochetes, still moving, ultimately became movers of chromosomes. Mitosis evolved. Spirochete attachments became centriole-kinetosomes. Perhaps some of these structures—those that have retained the power to reproduce—still contain DNA.

Whatever the precise scenario for the acquisition of motility and the sometimes respiratory and photosynthetic talents of eukaryotes, symbiosis most assuredly belongs in the narrative. Intimate symbioses were essential to the evolution of cells.

Strange New Fruit

The hypothesis that former spirochetes and thermoplasmas merged to form swimming protists is under investigation now. These merged beings may have been the original members of confederacies of bacteria from which all larger life evolved. But what about other symbiotic bacteria? How did they become involved?

Think back to those blue-green, photosynthetic bacteria that polluted Earth with oxygen gas. After reacting to make new minerals such as sulfate (SO_4), magnetite (Fe_2O_3), and hematite (Fe_3O_4) all over the planet's surface, oxygen waste began to accumulate in the atmosphere. Newly appeared gaseous oxygen killed off untold hordes of organic beings. Even today certain kinds of cyanobacteria are sickened by their own oxygen; *Phormidium*, for example, lives only in muds near other organisms that can quickly use up the oxygen it produces in would-be fatal concentrations.

Early on, cells evolved a tolerance for oxygen at low concentrations. Many modern prokaryotes still function best at oxygen levels of about ten percent— half of the atmosphere's typical concentration today. Oxygen-tolerant bacteria produce enzymes such as catalases, peroxidases, and superoxide dismutases that react with the dangerous gas to produce innocuous organic compounds and water. Without such chemical buffers the carbon of organic tissue is scorched, torched, and laid to waste by oxygen.

Nonetheless, the mitochondria of our cells come from bacteria that neither shunned nor merely tolerated oxygen. The bacteria that evolved into the matrilineally transmitted mitochondria—only the ovum bequeaths them to the human embryo— exploited oxygen's great reactivity. Like nuclear physicists devising a way to power spacecraft by using environmentally hazardous plutonium, the mitochondrial ancestors turned an intense danger into a radical opportunity.

In perhaps the greatest example ever of recycling, bacteria employed reactive oxygen to improve cell processes of energy transformation. Oxidizing the material they produced by trapping the energy of light, purple photosynthetic bacteria increased their ability to metabolize ATP, the energy storage compound—the biochemical "coin" used by every cell of every living being. Breaking down organic molecules and producing carbon dioxide and water, bacteria diverted the natural combustion of oxygen to their own purposes. Whereas on average two molecules of ATP are produced by fermentation of a sugar molecule, with the evolution of respiration the same sugar molecule was made to yield as many as thirty-six ATP molecules. The new bacteria— including the ancestors of our mitochondria— recouped energy from sugar molecules with over fifteen times the efficiency of their oxygen-poisoned predecessors.

That our mitochondria's ancestors were oxygen-respiring purple bacteria has been shown beyond a doubt by DNA sequencing. Like a village ransacked by barbarians that ultimately became civilized, fermenting organisms were attacked by oxygen-using predators that became mitochondrial laborers. We suspect the earliest hosts were *Thermoplasma*-like archaebacteria (already squirming with spirochete symbionts), able to withstand heat and acid but not free oxygen. These consortia evolved into the first protists, their spirochetes becoming undulipodia. The *Thermoplasma* lineage is implicated in this major evolutionary event because modern represen-tatives resemble the nucleocytoplasm portions of eukaryotic cells. *Thermoplasma acidophilum*, for example, possesses histone-like proteins nearly universal in larger life forms—animals, plants, fungi, and some protoctists—but lacking in other prokaryotes. The presence of histone proteins in human chromosomes may be a direct inheritance from the protists that were invaded by proto-mitochondria.

The invaders were probably of the "purple bacterial lineage," as classified by Carl Woese. These proto-mitochondria may have been similar to modern oxygen-respiring, rod-shaped bacteria such as *Paracoccus denitrificans*. This bacterium contains more than forty enzymes in common with human

Bdellovibrio, a predatory bacterium. Phylum: Pseudomonads. Kingdom: Monera. *Bdellovibrio* is caught in the act of invading a larger bacterium. It will multiply inside its prey, devouring away until it breaks the corpse in a living explosion, releasing a new generation of predators. Such invasions when they became permanent, in the sense that the smaller bacteria survive and multiply inside the larger bacteria, could have been analogous to the evolution of cells with mitochondria.

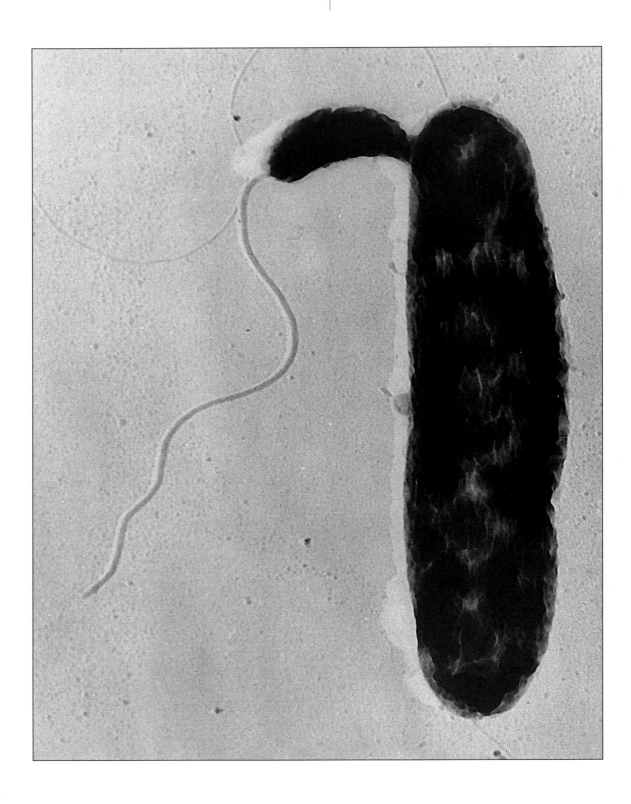

WHAT IS LIFE?

mitochondria. More probably, they were similar to the respiring *Bdellovibrio* or *Daptobacter*—modern predatory prokaryotes in the habit of attacking and multiplying inside larger bacteria. [PLATE 36] Eventually, the victims explode and a battalion of intruders merrily swim out. *Daptobacter, Bdellovibrio*, and similar unnamed bacteria are necrobes— beings that live off the death of others. But even if they began as a parasitic infection, the ancestors to the mitochondria did not stay that way. Fed and protected in a living environment, the proto-mitochondria were better off not destroying their oxygen-intolerant hosts.

Today, although mitochondria still possess their own DNA and still reproduce like bacteria, they cannot live on their own. The parasitism has become permanent: neither partner can escape, neither can survive separation. The first protists were thus odd couples, the results of fusion of two, or (in the case of plants), at least three once-independent beings. But unlike the fire-breathing being of Greek mythology who has the head of a lioness, the midsection of a goat, and a dragon's tail, these chimeras were real.

How do predators become symbionts? How does a deadly infection become a bodily part?

The Korean-American biologist Kwang Jeon at the University of Tennessee has already witnessed such a transformation in the laboratory. The answer is thus less a mystery than before. Jeon's experiments dramatically show how bacteria can change from virulent pathogens to needed organelles.

Like many of science's most amazing discoveries, Jeon's came about accidentally to the prepared mind. To his initial dismay he found one day that his amebas, which he grew in laboratory dishes, were sick and dying. Microscopic investigation revealed that each *Amoeba proteus* was infected with some 150,000 strange bacteria. All but a few amebas died. Curious about the moribund survivors, Jeon inject-ed new, healthy amebas with infectious bacteria

taken from the moribund. Most newly injected amebas died within a few days although, again, some managed to survive. Those that did repro-duced more slowly. After some months all the sur-vivors were infected. But these survivors had fewer bacteria inside them than those which had died.

After growing generation upon generation of infected amebas, Jeon extracted the nuclei from several. He transplanted these nuclei into healthy, bacteria-free amebas, whose own nuclei had been microsurgically removed. The amebas with the transplanted nuclei died on the third or fourth day—unless Jeon rescued them with a needleful of bacterial "infection." The disease had thus become the cure. A deadly bacterium had become a vital cell part.

Decades later Jeon's infected amebas are alive and well and living in Knoxville, Tennessee. His experiments have been repeated numerous times, and now he observes that the amebas differ in many features from their never-infected ancestors. Pathogens have become symbionts on at least four occasions. Symbionts have become organelles each time. Invader and invaded merge, evolve into new life forms. Branches on the tree of life do not always diverge but sometimes come together to produce strange new fruit.

Wallin's Symbionts

In 1927 the American biologist Ivan Wallin (1883-1969) wrote, "It is a rather startling proposal that bacteria, the organisms which are popularly associ-ated with disease, may represent the fundamental causative factor in the origin of species."[5] He claimed to have grown mitochondria outside their animal "host cells." Publicly shouted down by colleagues, Wallin gave up defense of the bacterial origin of mitochondria while still in his forties.

Wallin was almost certainly mistaken, as no one has ever been able to grow mitochondria by them-selves. Nonetheless, Wallin's theoretical assertions

Chlamydomonas nivalis,
an alga. Phylum: Chloro-
phyta (green algae). Kingdom: Protoctista. Red
snow algae in Antarctica.
The green chlorophyll
is masked by the red
pigment. DNA studies of
the green algae and plants
and marine red algae point
to a symbiotic origin of
the colored cell parts from
cyanobacteria. Cyanobac-
teria, teaming up with
larger cells, eventually
evolved into the plastids
of all "higher" photosyn-
thetic beings from
seaweeds to maple trees.

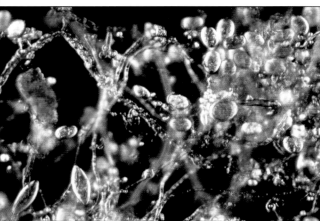

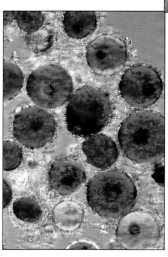

38 (middle)
Medium range photo-
graph of the snow algae
Chloromonas sp. and
filamentous fungi.
The pigments of these
photosynthetic creatures
gives the mountainous
snow its orange tint.

39 (bottom)
Microscopic view of red
snow alga *C. nivalis,* so-
called "watermelon or
candy-colored snow."
The red carotenoid pig-
mentation network serves
as a photoprotectant from
bright sunlight. The cells
are 400 times smaller
than they appear in this
photograph.

were prescient. Plant and animal life, he asserted, had appeared through what he called "symbionticism" or "the formation of microsymbiotic complexes." He meant new species form by the permanent acquisition of symbiotic bacteria.

Today Wallin has been vindicated. His 1927 classic book, *Symbionticism and the Origin of the Species*, was the first systematic description in English of the importance of symbiosis in cell evolution. Although heresy only decades ago, contemporary biologists agree that animals, fungi, and plants evolved from protoctist ancestors themselves originating from symbiotic bacterial associations.

The crucial piece of evidence unavailable to Wallin until just before he died was the discovery that mitochondria and plastids possess their own DNA. Wallin knew though that mitochondria and plastids tend to reproduce at different times than do the cells in which they reside — as if demonstrating a residual impulse of their earlier, wilder days. Bacterial respirers, like those infecting Jeon's amebas, allied with nucleated swimmers to form the ameba-like ancestors of larger life forms: aerobic protists. Combining metabolism and genes, different lineages of aerobic protists went on to evolve into animals and fungi.

Algae and plants are a further chapter in the same story. In subsequent symbiotic events, swimming protists that had already fully integrated with purple bacteria (now mitochondria) came to possess plastids. How? By indigestion. The resistant green bacteria — the food — remained alive inside transparent, vegetarian protists. A continuous supply of photosynthate — food made by the trapped photosynthetic bacteria — rewarded the protist, which quickly developed a penchant for sunlit waters. Like small farmers who harvest their own gardens rather than shop at a grocery, protists which incorporated their captives became increasingly self-sufficient. In return for the favor of food, the engulfed photosynthetic bacteria received a place to live and rapid free transportation into the sunlight.

These swimming protists, which later evolved into algae, were living greenhouses. Would-be food, really endosymbiotic bacteria, photosynthesized inside the luxury prison of live cells. The original undigested food was probably similar to *Prochloron*. This grass-green bacterium grows in the rear chamber — the cloaca — of homely marine creatures known as didemnids or "sea lemons." [see PLATE 9] *Prochloron*-like bacteria are a good scientific choice for the plastids of algal and plant cells. Spherical prochlorons and a rod-shaped (but similarly grass-green) bacterium called *Prochlorothrix* make precisely the same pigments — chlorophylls *a* and *b* — made by green algae and plants.

Multi-tentacled hydras, relatives of jellyfish and coral, are white but tint green when they possess symbiotic green photosynthetic microbes. The snail *Plachobranchus* has gardenlike rows of green plastids under its parapodial folds, part of the digestive tract. The giant clam, *Tridacna*, hosts green dinomastigote algae. Many organisms have allied with photosynthetic bacteria or algae. History repeats itself.

Grass-green and blue-green bacteria are independent versions of the plastids of algal and plant cells. Algal plastids need not be green. Plastids of the alga responsible for the red tint of alpine snow patches in the late spring and summer ("watermelon snow") are red. [PLATES 37, 38, and 39] And in Tanzania's Lake Natrum swoop great flocks of pink flamingos. Red photosynthetic bacteria and algae with red plastids, pigmented with the same carotenoids that color carrots, grow in the lake. Flamingos look pink because the pigments at the microbial base of the food chain wind up coloring the bodies of these intriguing birds.

Genetic evidence, DNA, RNA, and protein sequence information links red algal plastids to certain cyanobacteria with the same forensic accuracy admissible in court to convict a rapist whose DNA matches that of a sperm sample. The multicolored bacteria of the Archean eon have not gone away. They have joined with other cells to become the sea-green chloroplasts of garden cucumbers. Others have become the brown phaeoplasts of kelp in coastal waters. Still others lurk today as the red rhodoplasts of dulse, a form of sea lettuce. If food crops are grown in orbit, on Mars, or on other planets greened with life, it will be a transhuman phenomenon, part of the same bacterial expansion that began more than 3,000 million years ago on the Archean shores.

Multicellularity and Programmed Death

Plants and animals are so complex that it is easy to forget their original status as colonies of hybrids. Occasionally, however, we are reminded of our multicellularity. "HeLa" cells—from the cervix of Henrietta Lane, a woman who lived in Washington, D.C.—continue to be grown in laboratories around the world, despite Lane's death from cancer of that same cervix in the 1950s. This morbid medical fact demonstrates our colonial nature as huge collections of nucleated cells organized into tissues.

By symbiosis different varieties of bacteria came together and made cells with nuclei. These cells with nuclei often cloned themselves into multiple copies that stayed in physical contact after reproduction. A *Paramecium* or *Euglena* is an "individual" nucleated cell, already fascinating in its mixture of living beings. But plant, animal, and fungal life greatly expanded the complexity of the free-living protist cell by repeating it to make multicellular copies that ultimately evolved into separate tissues, such as reproductive and nerve tissue, with distinct functions.

The offspring of some of these protists, starting in earnest perhaps a thousand million years ago, failed to separate after they reproduced by cell division. They began permuting themselves into colonies, some of whose members died each generation. Thus certain colonial protists became

physically large members of the group, and the diversity of protoctists evolved. Looking at modern protoctists suggests how such colonies could have formed from individual cells. Animals, including of course ourselves, are transformed colonies of protist cells.

Charles Darwin emphasized that evolution occurs as different individuals pass on their traits by out-reproducing others. But individuality, always in flux, is relative. Cells form and interact in a wide array of configurations. Together they form individuals at various size levels and degrees of interdependence. The alga *Chlamydomonas*, with its large green single chloroplast, is a bacterial composite. *Volvox*, a spherical confederacy of *Chlamydomonas*-like protist cells, is a green multicellular descendant of *Chlamydomonas*, just as animals are multicellular descendants of swimming protists. [PLATE 40]

The origin of any "individual" large organic being depends on integrative gene-transferring processes not easily reversed. These integrative processes first stabilized as the colonial protoctists evolved from free-living protists. *Volvox* algae, like other protoctists, fungi, plants, and animals (but unlike bacteria), do not casually trade their genes. Larger organisms simply cannot trade genes the way bacteria do.

Any single protoctist, plant, fungus, or animal is a member of a species. Most likely, protoctists were the first organic beings to form species and the first whose species went extinct. The origin of individuals who all belong to the same species is identical to the origin of the first protoctists. Canadian microbiologist Sorin Sonea makes a good point when he claims that bacteria, because on a planetary scale they reversibly trade genes, do not have true species. Species are groups whose members interbreed. All bacteria on the planet can in principle interbreed. If anything, they might be said to form a single, global species.

Species demarcation is thus much more applicable to the protoctists, in which, indeed, it first

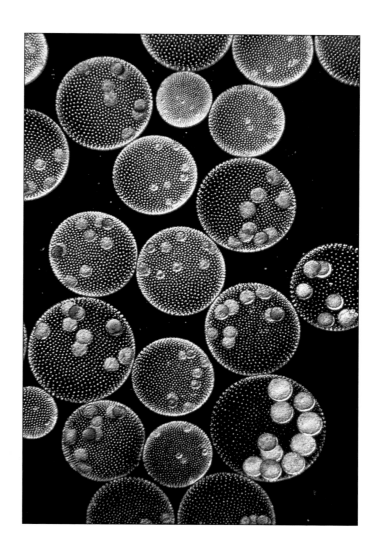

40

Colonies of *Volvox*.
Phylum: Chlorophyta
(green algae). Kingdom:
Protoctista. Individual cells
of this green colonial alga
resemble free-living cells
of *Chlamydomonas*. The
evolutionary move from
unicellular to multicellular
"individuality" is a crucial
one that has occurred
many times. It may be happening again as electro-
nically communicating,
technologically interacting
human beings form net-
works required for survival.
(See chapter 9.)

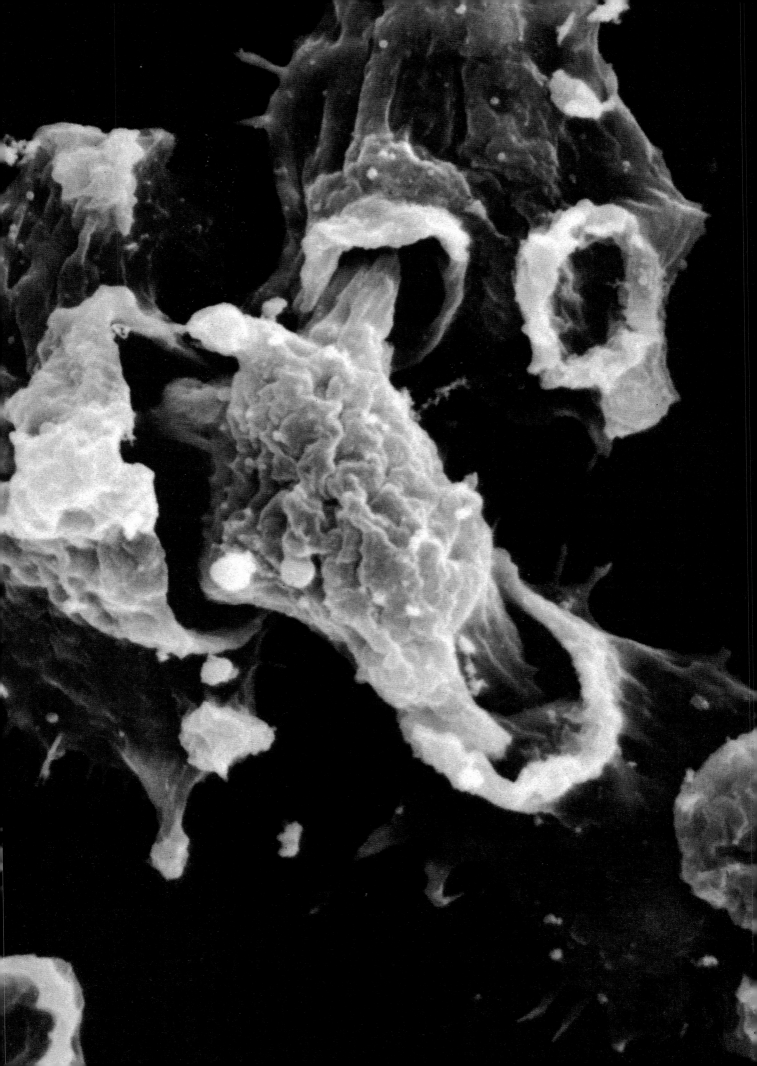

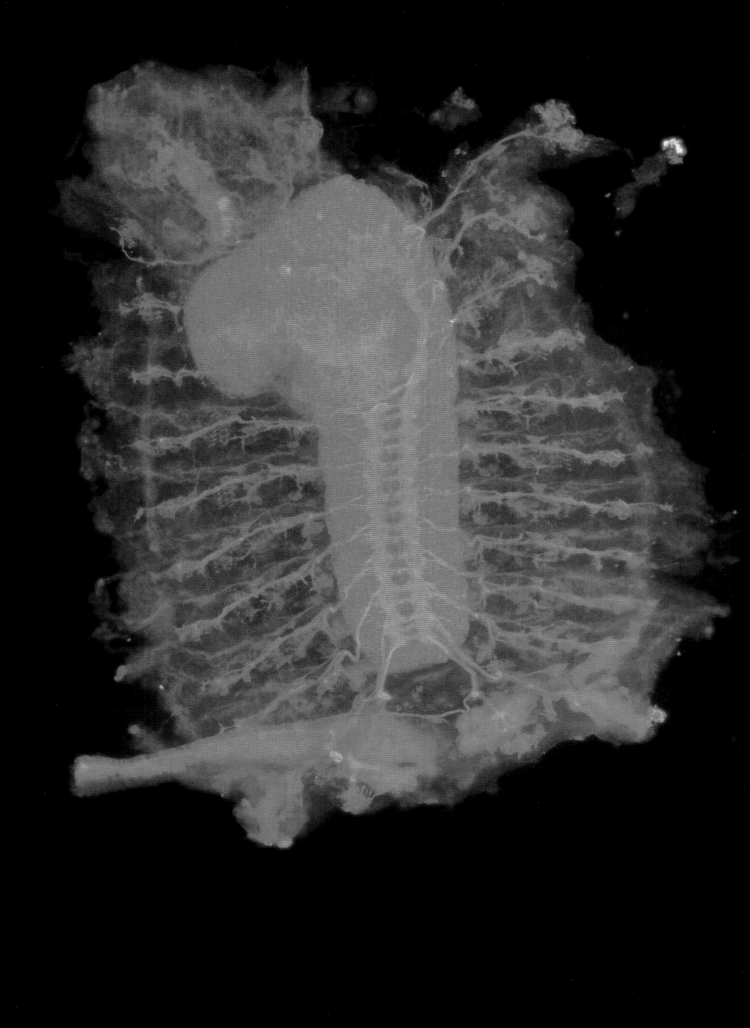

Composed of tissues with distinct connections between the cells—desmosomes, gap junctions, septate junctions, and the like—the bodies of animals are individualized. Such cell-to-cell connections, unknown in any other kingdom, must be produced by embryo development. In a predictable sequence, cells of the embryonic animal divide by mitosis, they roll over one another and set up alliances such that many, sometimes most, of the animal's body cells must die in a preprogrammed fashion. If the young embryo cells do not die on cue or do not establish definite connections to send specific signals through their junctions, no animal body develops. Animal embryos are crucial.

In animals the muscles, nerves, and circulating fluids such as blood all derive by mitotic cell division and differentiation from the blastula. Doomed to mortality, the blood, muscle, and nerve cells stop dividing; their programmed death contributes to the construction of the body and the continuing of the theoretically immortal egg and sperm. Without the blastula embryo formed from the fertile union of sperm and egg, animals would not exist. The blastula embryo is the animal universal in all thirty-three or so phyla. In the usual course of animal development the cells of the blastula continue to divide, move, and die. They form the next stage, the gastrula, which makes a new mouth at the front end of a digestive tube. The mouth is succeeded by a swelling, a stomach, and an anus in this unfolding process, called, logically enough, "gastrulation."

The vast majority of blastulas in the thirty million animal species "gastrulate" and end up with a distinctive "tube-within-a tube" digestive system, committed to heterotrophic nutrition. Formed by cell-to-cell communication via gap or septate junctions with nerve cell synapses, the tube or gut establishes the ingestive nutrition so nearly universal in the kingdom. But a few kinds of exceptional animals lack intestines and even these must pass through a blastula stage. This kind of embryo is the hallmark of any animal.

Neither fungi nor protoctist, although most are heterotrophs, forms any embryo at all. Probably correlated with the absence of the embryo is the absence of animal-style distinctive individuality. Plant individuality is far less fixed than that of animals. Although all members of the plant kingdom do form embryos, plant embryos are very different from any blastula. Each plant cell is walled from its neighbor, precluding the movements and realignments that all blastulas undertake as they become individual animal larvae or adults. An embryonic plant cell cannot form gap or septate junctions, nerve synapses, or any other of the animal cell-to-cell connections. Stationary in its place, the plant cell only grows, dividing by mitosis, or dies. It is the fateful blastula that presages all the nuances of animal behavior and distinguishes our kingdom from all the rest of life.

Great Grandparent *Trichoplax*

To the bacterial realm goes the award for metabolic innovation. As keepers of the biosphere, prokaryotes are the most inventive forms of life, and their descendants include the now-essential organelles within our own cells. The protoctists, too, dealt originally with the problem of environmental threat: in autopoietically changing to stay the same, they arrived at death, sex, and metamorphosis into resistant structures. But with the origin of animals nature seems to have reached new levels of playfulness, awareness, complexity of form, responsiveness, and deception.

A butterfly's wing bearing an imitation raindrop with a line displaced just as if it had been refracted through real water; a cheetah poised to pounce; an acrobat juggling upon a high wire: animals amaze.

Today's minimal animal is *Trichoplax*—a headless, tailless creature discovered crawling on its belly along the side of a marine aquarium in Philadelphia in 1965. Were it not for its sex life and its embryo,

Trichoplax would be a protoctist. Wafted along by its undulipodia, *Trichoplax* superficially resembles a slime mold slug or a giant ameba. But it is a multicellular being throughout its life and is a true animal. It has more undulipodia on its belly than its back. (Here it resembles an insect, which has more undulipodia on its legs than its antennae.) Having neither head nor hind end, right nor left side, no eyes, no stomach, this minuscule slow crawler gives away the secret of its animality only at reproduction. After fusion with sperm, a spherical *Trichoplax* egg becomes a blastula embryo that, with more cell division, flattens and, ameba-like, slinks away. Although you probably wouldn't want to hang a portrait of it in the drawing room, *Trichoplax* likely bears a strong resemblance to our earliest animal ancestors.

A sponge is an animal composed of only a few types of functionally and morphologically different cells. For example, those on the outside may grow glassy rods for support and protection; those on the inside use their undulipodia to maintain a flow of water from which food can be extracted. If yellow and orange (*Haliciona*) sponges are squeezed through cheesecloth so that each is broken into bits and mixed with the other, the cells manage to find their own kind in the aqueous environment. After a few hours the cells reorganize into fully formed and distinct yellow and orange sponges. So too, a freshwater polyp, a jellyfish relative having about a hundred thousand cells of a dozen cell types, can be disassociated into single cells. In permissive solutions they begin to rearrange themselves. Unlike sponges, they cannot complete the process. Monstrous growths, in which head, gut, and foot (basal stalk) realign unsuccessfully, are the result. In this case, the integrating mechanisms assuring autopoietic self-maintenance fail.

In most colonial green algae and ciliates (all of which are protoctists) any single cell may separate and reproduce on its own. In others, only certain cells reproduce. The theme of animal evolution, the development of discrete individuals, involves curtailing reproduction in favor of specialization. Protoctist anarchies, in which any cell could reproduce, were replaced during the emergence of animals by cell oligarchies, in which only a few (sometimes a very few) had the privilege of living on into the next generation by way of progeny.

The transition from cell, to cell society, to animal organism is an old story in evolution: individuals group into societies, which themselves become individuals. Under intense selection pressures, swimming protists became colonial protoctists. Then, in the later Proterozoic eon, *Trichoplax*-like animal bodies appeared. The specialization of massive numbers of cells into integrated individuals is at the base of animal life—and of those later groups, fungi and plants.

Sex and Death

Only accidental, externally caused death, existed at the origin of life. So it was for a long time thereafter. But with protoctists came "programmed death": death in which cells age and die as part of the life of the individual. In familiar animals—insects, mammals, and birds—the difference between the part that dies and the part that potentially lives on is the difference between the body and the sex cells. In mammals the sex cells (or "germ plasm," as biologists sometimes say) are the only cells whose direct progeny survive into the next generation. In contrast to the ova and sperm, the "soma"—the animal body—has a discrete life span.

With a high degree of precision, animal cells must reproduce—or cease reproducing. For example, during the intrauterine development of the mammal brain more than ninety percent of the cells that develop die before the fetus becomes an infant. These brain cells stop growing and disintegrate, are sacrificed in the process of growing a healthy infant. The essential difference between the living germ cells and the dying body cells of animals is likely very old.

We speculate that the ancestors of animals were composed of relatively few cells that differentiated into at least two distinct kinds. One kind specialized in using their 9(2)+2 microtubule organelles to form undulipodia for propulsion, for sensing prey, for fostering water flow over or through the animal, or for sweeping food particles into and along digestive systems. [PLATE 46] But it is an oddity of physiology that once animal cells dedicate their centrioles to forming undulipodial shafts they can no longer use them to create the motility apparatus for mitotic cell division. This means that animal cells stood to gain by sticking together in specialized colonies. Even today an animal cell, whether of a tissue or a sperm after growing the undulipodium, no longer divides. A centriole forms a kinetosome and relinquishes cell immortality; a kinetosome cannot revert to a centriole. The irreversibility of kinetosome formation appears to be an inviolable rule within the animal kingdom. Animal cells can either form kinetosomes (grow undulipodia from centrioles) or reproduce by mitosis—but not both. An animal cell with a kinetosome is a dead animal cell—its days are numbered, as it will not divide again.

Perhaps the DNA reported by David Luck and John Hall to be in the kinetosome-centriole is used for mitosis or to form an undulipodium but not for both. Like choking from inhaling water, any attempts of cells to simultaneously reproduce and maintain undulipodia would have been thwarted. And yet animals seem to have found an answer to this genetic dilemma: by sticking together in colonies—colonies where some cells reproduce while others form undulipodia—they could in effect have their cake and eat it too. The restriction of a cell unable to divide after growth of its 9(2)+2 organelle was overcome by colony formation. The great majority of cells retained their option to divide, while a few sacrificed immortality to be

46
Undulipodium in cross section. The shaft (axoneme) displays the 9(2)+2 arrangement of microtubules. This distinct intracellular organization is found in widely diverse beings throughout the natural world from the sperm cells of men to those of ginkgo trees. Electron micrographs of cuts through the shafts of the cilia propelling swimming paramecia and trichomonads and the cilia that push the egg through a woman's Fallopian tube also reveal this 9(2)+2 pattern.

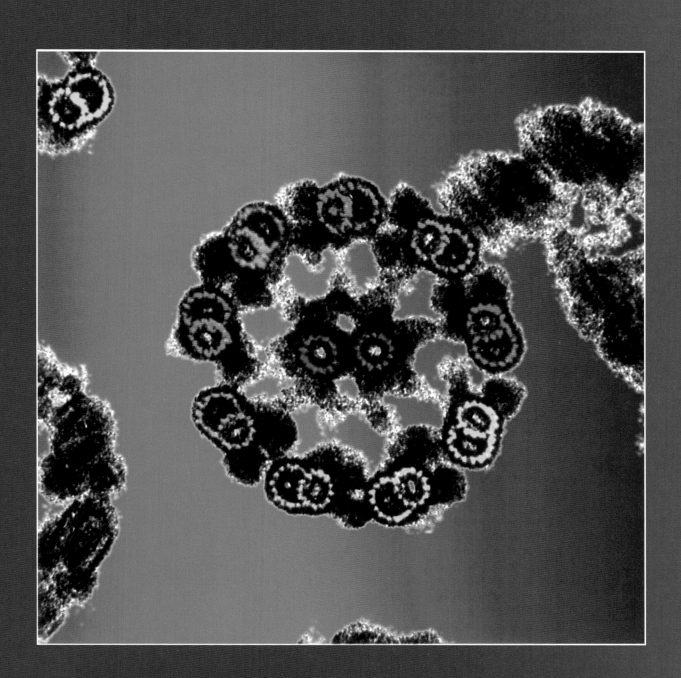

undulipodiated. But even the cells that do divide in the animal do not do so indefinitely. After 600 million years the adult animal is still a mated protist's way of making other mating protists.

Our whole life from womb to tomb is in fact an interim stage in the life cycle of tiny fused cells. Animals emerge into another dimension, visible life and consciousness, only to return via sex to their ancient single-celled, microbial state. Death is the price we all pay for this ancient history of multicellular compounding, for this inability of hungry protists to undo their Proterozoic entanglements. What "dies" is the body, the adult flesh after it has released into the water or body fluid the protist-like tailed sperm and chubbier egg. Animal life did not appear de novo, but from protoctist predecessors. Protoctists with elaborate cycles of fertilization, multicellularity, and meiosis became animals.

Like programmed death, gender is not intrinsic to life. Gender evolved. Cells of different mating types, like protoctist lovers today, were initially identical to each other in appearance. The seasonal merging and restoration of chromosomal numbers in fertilization set the stage for the origin of gender. The first mates met slapdash in a watery environment then, as protoctists do today. Responding to slight chemical differences in each other, mates came together. Sponge, sea urchin, fish, and even mammalian sex cells, like their protoctist ancestors, still meet in watery places.

Animal cells continue their ancient practice of aquatic encounter. The sex cells of oysters and even some frogs and fish meet directly in the water, to fuse unattended by adult bodies. In reptiles, birds, and mammals, however, sexual mergers occur *in vivo*. Genitals evolved independently in many animal lineages. The penis or intromittent organ of the male created a delivery system for sperm. The female genital tract afforded the ova a protected place where fusion could occur. The many small sperm of males compared to the few larger eggs of females was the beginning of an evolutionary

asymmetry which today expands into the realms of political, sociolinguistic, and psychological debate. Evolutionary biologists suggest that early sexual inequality—males maximize reproduction by inseminating the largest possible number of females, whereas after a certain limit mating becomes superfluous to females constrained by devotion to their lesser quantity of eggs—is behind distinct male and female attitudes toward sex.

Cambrian Chauvinism

English geologist Adam Sedgwick (1785-1873) named the time period to which the oldest fossils belonged Cambrian, after "Cambria," the old name for Wales in southwestern Great Britain. To him and other early paleontologists the appearance upon Earth of animals seemed miraculously sudden. All prehistory prior to Sedgwick's Cambrian became known as the "preCambrian." Until the late twentieth century the origin of the Cambrian animal fossils was considered "the most vexing riddle in paleontology."[2] So quick was the apparent appearance of animal life in the fossil record—not only in Wales but also in Newfoundland, Siberia, China, and the Grand Canyon of Arizona—that it is still referred to as the "Cambrian explosion."

Today, much of the answer to the riddle is known. Foraminifera and other conspicuous protoctists, the "protozoa" whose fossils were once dismissed as those of tiny invertebrate animals, in fact preceded animals by at least 500 million years. [PLATE 47] Because, like early animals, most protoctists were small and did not form hard parts, they remained largely undetected and unpreserved. Life prior to the Cambrian, despite its astounding biochemical and metabolic innovations, is still often dismissed as "preCambrian"—with the connotation that nothing much worth mentioning in evolution happened between the origin of life and the appearance of shelled animals.

Bacteria and protoctists set the stage. They, not

48
These fossils called sclerites appear at the beginning of the so-called Cambrian explosion. Whether they are fossil fragments of the very first animals or protoctist predecessors to the animals has not been established.

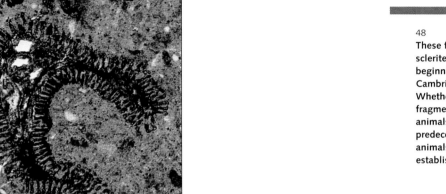

47
Giant acritarch, an unidentified fossil microbe, nearly a millimeter in size. This acritarch is found with the mineral phosphatite. It is suspected that the beings which fossilized as acritarchs were among the oldest protist cells. Animals did not appear *ex nihilo* but from colonial protoctists in soft-bodied organisms that left scarce traces in the fossil record. The Cambrian "explosion" had a long microbial fuse.

animals, introduced DNA recombination, locomotion, reproduction leading to exponential growth, photosynthesis, boil-proof spores. They, not animals, pioneered symbiosis and the organization of individuals from multicellular collectives. They invented intracellular motility (including mitosis), complex developmental cycles, meiosis, sexual fusion, individuality, and programmed death. The prokaryotic microbes, not animals or plants, still run all the geochemical cycles that make the planet habitable. The protoctists, in their new status as individuals from coevolved bacterial communities, invented resistant cysts, skeletons and shells, gender behaviors, cell-to-cell communication, lethal toxins, and many other processes later co-opted by animals. Animals were preceded by bacteria and protoctists, not by chemicals. The animal explosion had a long microbial fuse.

Patterned fossils of tiny shelly plates, known collectively as sclerites, mark the beginning of the Cambrian about 570 million years ago. [PLATE 48] The lowermost time-rock division or period of the Phanerozoic eon is the Cambrian. Distinctive animal fossils abound in Phanerozoic rocks overlaying barren Proterozoic strata. Sedimentary rocks from all over the world deposited 530 million years ago (still in the Cambrian) contain a striking array of skeletalized marine animals. At about this time, brachiopods (lampshells) and annelid worms appeared. So did trilobites and other arthropods (of which insects and lobsters are modern examples).

Some paleontologists still wonder how these various phyla could have cropped up "all of the sudden." Noting iron rust and other clues that oxygen had entered the atmosphere 2,000 million years ago, some scientists suggest that a threshold level of atmospheric oxygen (O_2) itself induced animal evolution. But any scenario to account for the "sudden" appearance of animals is almost surely a misreading of the evidence. Animals, although they evolved late in the history of life, did not evolve suddenly. "Seemingly," writes paleontologist Harry B. Whittington, "there was a long period of metazoan [animal] evolution before the Cambrian, but it is only in the earliest Cambrian rocks that minute shells of metazoans appear.... The Burgess Shale shows that it was not only metazoans with hard parts that were diversifying in the Cambrian, but also soft-bodied metazoans, including coelenterates, worms, arthropods, chordates, and various strange animals."[3]

The Burgess shale is a collection of Cambrian fossils exposed on a high mountain in Yoho National Park in British Columbia, Canada. Discovered by Charles Walcott in 1909, the exquisite and numerous Burgess shale fossils have given paleontologists lifetimes of work. Because even soft-bodied animals were preserved in it, this shale is a treasure. Shallow marine dwellers were preserved in the underwater mudslides that made the Burgess shale. These were a large variety of organisms, some similar to modern forms, others with no known descendants. Among the beautiful, if monstrous, animals are *Opabinia*, a five-eyed sea-bottom crawler with curved tail fins and a grasping, jointed organ that suggests it was a formidable predator, though only four inches long.

Hallucinogenia, in accordance with its name, has puzzled paleontologists because until recently no one was sure which side was up (spikes as armor) and which side was down (spikes as legs). Of the many Cambrian arthropods that the Burgess shales have preserved, only one sort gave rise to a lineage that much later evolved into the vast array of land creatures with six legs known today as insects. Had evolution taken another course, another Cambrian arthropod—or an entirely different animal, for that matter—might have gone on to populate the continents.

Among the most touching of the Burgess specimens is *Pikaia*, the first known member of our own chordate phylum—the phylum to which humans and all other animals with backbones belong. *Pikaia*, a segmented wormlike swimming creature, is inconspicuous compared to more spectacular Burgess forms. But it had a solid cartilaginous rod—the notochord—running down its back. This universal structure of chordates if not present in the adult is fleetingly apparent in larvae or other immature life cycle stages. Until the discovery of *Pikaia* in the Burgess shale, no chordates were known from any rock older than about 450 million years, deposited during the Ordovician, the geological period after the Cambrian.

The Burgess chordate is a stunning discovery because it shows that the predecessors of darting *Pikaia*—which may have been the ultimate ancestor of all fish, amphibians, reptiles, birds, mammals, and us—were alive and navigating the muddy waters 530 million years ago. The success of *Pikaia* may be directly responsible for the later emergence of such a wide diversity of forms, from mugwump to turtle, moose, rabbit, and giraffe. The presence of streamlined *Pikaia*, with its pointy, somewhat cobralike flattened head and bifurcating snail-shaped tail, shows that our ancestors swam in the primeval oceans.

Before the armor-headed trilobites crawled the planet, before droves of clam-like lampshells expired in Cambrian muds, before eurypterid "sea scorpions" left their hard exoskeletons in the fossil record, soft-bodied animals proliferated. Even less obvious and far older than the Burgess shale animals are the "Ediacaran" beings preserved in sandstones 700 million years old—before the Cambrian period, before the Phanerozoic eon. Most are probably not animals at all but bizarre, extinct protoctists. In the 1950s Martin Glaessner of the University of Adelaide named these extraordinary fossils after a rock formation in the Ediacara Hills of Australia. Similar soft-bodied beings have been found in England, Namibia, Greenland, Siberia, and some twenty other localities.

The Ediacaran organisms seem to have been floating gelatinous beings enjoying shallow water at sandy beaches. Some were flat, others "quilted," others intricately textured organisms. They ranged, shapewise, from leaflike *Pteridinium* to three-armed *Tribrachidium*. But these Ediacaran beings seem to have formed no hard parts, eggs, sperm, or blastula embryos. They may have been large protoctists, animals, or both. Some of the larger Ediacaran beings probably photosynthesized in shallow coastal seas. Others fed on bacterial pastures. But their lack of armor indicates that large predatory organisms had not yet evolved—it was truly a "Garden of Ediacara."[4]

Ediacarans may have been ancestors to the Burgess Cambrian animals or, more likely, because they are utterly unique, they may have been one of evolution's many "false starts." The earliest marine animals—whatever these were—may have fed on protoctists, including algae. The small size and relative mobility of these algae eaters probably guaranteed a nutritional niche without much competition. Only after animals began trying to eat one another—and evolving larger size and hard bodies

as defenses—did they become obvious in the fossil record. Mating, embryo-forming animals must have exploited resources for millions of years before they evolved hard, easily preserved parts. Cambrian fossils are only the tip of the iceberg of animal evolution. They were hardly "the first animals."

A thermodynamic truth is that as heat dissipates life organizes and its surroundings degrade. There is no life without waste, exudate, pollution. In the prodigality of its spreading, life inevitably threatens itself with potentially fatal messes that prompt further evolution. But sometimes waste can be fashioned into something useful.

Within protoctist or animal cells the concentration of calcium ions—Ca^{++} charged particles—is some ten thousand times less than that of sea water. Calcium phosphate becomes a fatal precipitate as "rocks" form inside cells when too much calcium enters. And by binding with phosphorus (as phosphate), errant atoms of calcium deprive the cell of an essential ingredient for making DNA, RNA, and membranes. By contrast, in controlled, small quantities calcium ions can be an intracellular resource. Poison depends on dose. In small doses calcium ions signal; indeed, they are part of the electrochemistry of thought. But the calcium excess must always be shunted outside of the cell. If the process of extruding calcium beyond cell frontiers lapses, then chemistry takes over. As people with kidney stones know only too well, calcium phosphate, where it doesn't belong, causes trouble. While animals remained soft they released calcium into seawater. Near the beginning of the Cambrian period in the Phanerozoic eon, however, some began to control their calcium extrusion.

Evolving, early animals converted potentially threatening blockages into living architecture. Our bones and skulls, like those of the fish-headed amphibians that preceded us, are still made of calcium phosphate salts (Ca_2PO_4). Some animals, such as corals, fashioned calcium, phosphates, and carbonates ($CaCO_3$) as their exteriors. Other organisms deposited calcium inside as teeth. Replacing organic cartilage, hard calcium phosphate infiltrated the proteins to provide the structural framework (shells and bones) attached to muscles. As armor emerged in some Cambrian creatures, teeth and poking appendages evolved in others to penetrate that armor.

Human industry has no monopoly on hazardous waste. Earlier life forms tell us by their examples that long-term survival involves not so much halting pollution as transforming pollutants. Termites build nests from feces and saliva. Pollution in the form of calcium excrement, patched and reworked by the busy muscles of animals, formed the basis for the first shells.

Evolutionary Exuberance

Becoming numerous, using their powers of movement to colonize new territories, animals evolved, thus creating and inhabiting new ecological niches. The Russian-American novelist and respected entomologist Vladimir Nabokov once suggested that patterns of certain butterflies look more like the whimsical touches of an artist god than traits that had merely blindly evolved. Yet with a fuller, less mechanical view of evolution such animal traits can be explained.

Evolution is no mechanical law but a complex of processes, sensitive and symbiogenetic, in part resulting from the choices and actions of evolving organic beings themselves. Natural selection is often said to "favor" this or that trait. But the nature that selects is largely alive. Nature is no black box but a kind of sentient symphony.

Of all the organisms conceived, spawned, hatched, and born, very few survive. Tastes in food and mates lead certain beings to generate more offspring than their fellows produce. In some cases females help designate the genetic makeup of populations by choosing the healthiest, most ostentatious, or strongest males. But conscious intervention in evolution by those so evolving can be of a more subtle variety. A butterfly's wing with an imitation raindrop showing a line displaced as if it had been refracted through water does not require a conscious author, and yet it could arise from consciousness. Beguiling artifices might arise from misperception by intelligent actors, for example, birds consistently mistaking an insect's wing pattern for a leaf. Nature is made partly in the image of mind.

Nabokov was right to say that the greatest enchantments in both art and nature involve deception. The element of surprise is the revelation that a given phenomenon of the environment was, until this moment, misinterpreted. Animals who experience surprise as a pleasure are likely to recognize camouflage and leave more offspring than are their less perspicacious brethren. Selection as nature, filled with live, sensitive beings is by no means blind. [PLATES 49 and 50]

Deception is very important in animal societies — so much so that some sociologists speculate that human technological intelligence is an evolutionary offshoot of "Machiavellian" social intelligence — the

49

An angler fish with
bioluminscent spots and
ribs. Phylum: Chordata.
Kingdom: Animalia.
The family (ceratioids) to
which this fish belongs
has bioluminescent
members more capable
than we of culturing pure
strains of glow-in-the-
dark marine vibrio
bacteria in their bodies.
These deep sea fish use
their bioluminescent
symbiotic organ to lure
unsuspecting potential
prey who mistake the
protruding appendage
for a small edible fish.

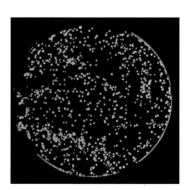

50

Photobacterium fischeri,
a bacterium. Phylum:
Omnibacteria. Kingdom:
Monera. Petri plate
showing colonies of lumi-
nescent bacteria. Many
types of fish cultivate
symbiotic bacteria in
special organs and put
their light to good use
in predator defense, food
illumination, or mate
signaling.

capacity to procure food, mates, child care, and so on by outwitting others in the tribe. The outwitting, outrunning, or outfighting need not be entirely conscious. Fearful apes and monkeys undergo a physiological reaction that stands their hairs on end. The effect to a potential combatant is an increase in size of the opponent, commanding if not fear then respect. The B-52, punk, and other "big hair" hair styles may similarly affect onlookers.

Since as naked primates we humans reveal more skin than hair, our own hair-raising response is rather measly: a prickly feeling, a hot flush across the nape of the neck, a tingling along the spine. Gooseflesh seems to be the evolutionary vestige of follicles that still go through the motions. Gooseflesh nevertheless exemplifies the evolutionary link between body and mind. A goosebump-mediated increase in a mammal's apparent size would be useless in a senseless world. But we live in a sensuous one, where details determine food and mate choices that in some cases spell the difference between life and death, between procreation and barrenness.

One of the sublime mysteries of life—supposedly calling evolution into question by its very presence—is the eye. Darwin wrote of the eye's "illimitable perfection." The eye, connected to the brain, perhaps seems perfect because it is the evolutionist's principle tool. But how might the eye, this subtle source of perspectival and reflexive mystery, have evolved?

The problem at first sight seems impossibly difficult. But not if we remember the microbes. Vision was anticipated in light-sensitive bacteria. Rhodopsin, "visual purple" of the mammalian retina, is a colored protein complex present in abundance in the pink, salt-loving archaebacterium *Halobacter*, where it is equally sensitive to light. The colored pigment portion of rhodopsin is retinal, a chemical similar to carotene of carrots and formed by the oxidation of vitamin A. Retinal, the absorber of light in the retina of the mammalian eye, has a 4,000-million-year history.

Using a plastid inherited from a cyanobacterial ancestor, the dinomastigote *Erythrodinium* functions as a kind of single cell eye. With its "imitation lens" and "imitation retina," this protist evolved a light-sensitive focusing device that involves most of its tiny body. Insects, flatworms, sea slugs, and frogs have eyes that are very different from one another, but all have carotene-derived light-sensitive membranes, lenses, and movable parts that direct light signals to locomotory organelles (such as undulipodia) or organs (such as muscles). Some evolutionists suspect eyes evolved in more than forty distinct lineages of animals. In all, light sensing is connected to movement in some way so that, once signaled, the creature can respond.

Sight, and its organ, the eye, may seem miraculous. Nonetheless, eyes exist along a continuum of complexity that both precedes mammals and is lately surpassing them, in forms such as infrared detectors, radio telescopes, and satellite imaging technology. Light sensitivity, in the rudimentary sense, even antedates life itself: colored compounds react in highly specific ways to visible solar radiation.

The human eye bears the marks of its microbial predecessors in other ways. The "rod" and "cone" cells are incapable of mitotic division; they bear kinetosomes and short undulipodia, inherited from their protist ancestors. The bony socket in which the eye resides derives from the autopoietic necessity of recycling calcium waste. Over time, life becomes more organized, integrating chemicals and even waste into beings so sensitive they eventually begin to perceive their own condition.

Messengers

By the late Devonian the phyla Arthropoda, Annelida, and Chordata had evolved representatives able to survive the rigors of life on land. Living beings evolved in water. The bacteria and protoctista cells since their inception were bathed in fresh and salty water. Desiccation was a dire threat any land pioneer had to overcome. Evolution of terrestrial species was no meager accomplishment. But the evolution of land animals was a triumph not just for individual organisms and species; it was a victory for the biosphere.

Movement and intelligence permitted land animals to act as vectors and messengers, to spread themselves to once-remote regions. By the early Tertiary birds had begun distributing phosphorus, a limited resource, to northern lakes and alpine peaks—simply by eating in one area and excreting in another. Carrying archaebacteria, ciliates, and other microbes in their rumens, cowlike animals digested grass, releasing methane, a greenhouse gas, into the atmosphere. Nitrogen-rich animal excrement accelerated algae growth, and fed fish and copepods in cold-water ecosystems. Especially during the Cenozoic, the most recent 65 million years, the quick reaction times, continent-crossing migrations, and complex social interactions of animals have accelerated activities within the biosphere.

But long before the Cenozoic, life was also a geological force. Photosynthesizing blue-green bacteria retained water in soil and sand, making Earth's surface green with chlorophyll. Carbon-precipitating life sequestered more and more carbon into coal and limestones, converting a tepid planet into one that indulged in glacial episodes. Land life created soil from planetary rubble. Ocean life transformed salts into reefs and evaporite flats.

"Proprioception" is the word for an animal's sensing various parts of its own body. Human beings today, likely the most populous of all mammalian species and certainly the most widespread, together behave as a kind of planetary proprioceptor, giving the biosphere sensations of itself. The greatest diversity of life exists in tropical jungles, such as the Amazon rain forest. Considering that distinct types of bacteria have merged to form the eukaryotic cell, and that colonies of eukaryotic cells evolved into animals, one wonders what may result from interaction of dense animal-rich communities.

Just as animal flesh was honed from the raw material of bacteria over eons, so complex interactions produced fledgling individuals at a scale beyond that of animals. Ants, termites, and bees form societies that carry out works in common. Reminiscent of human civilization, these arthropod workers methodically care for the young and divide labor among specialized castes of soldiers, workers, and reproductives. But whereas human civilization is only several thousand years old, fossil evidence shows that ants and bees have been organized into collectives for at least forty million, and termites for perhaps two hundred million years.

Together, animals confer their powers of movement and perception on the biosphere, making it an organized collective, the largest organic being of all. The animal actors of the global hive are at least 250 million years old. Snakes sense infrared radiation. Whales hear ultrasound. Bees detect the plane of polarization of visible light. Wasps see ultraviolet light patterns in flowers that look unpatterned to us. Dogs enjoy "ultrasmell." Sharks ferret out buried prey by detecting electrical potentials from the heartbeats of the hidden. Animals signal, and sense, and engage each other and their living environment in the visible, auditory, olfactory, and invisible radiative realms. Such sensitivity, so widely dispersed, sensitizes the entire biosphere.

Humans have extended one version of animal sensitivity into near Earth orbit. The image of Earth from space expands our awareness of the global environment. From the rudiments of animal sensibility and movement have come technological instrumentation, wheeled vehicles, and telecommunication. Together, the eyes of blackbirds, the sonar

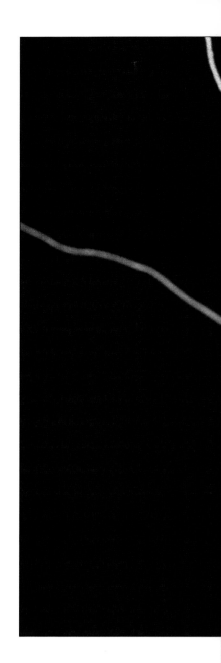

51
Eschiniscus blumi, a "water bear." Phylum: Tardigrada. Kingdom: Animalia. These microscopic animals, named water bears by English naturalist Thomas Huxley are known as tardigrades. They survive drying out in temperature ranges from 150°C to -270°C. These microbeasts occur all over the world, but because the largest are no more than 1.2 mm in length, they remain obscure. The span from claw to claw in the photo is less than 0.5 mm.

of bats, the heat absorption of worms, the bacteria-derived luminescence of marine fish, and the aggregate awareness of untold numbers of walking, crawling, flying, burrowing, thinking beings produce more than the sum of the parts. Sensitivities interact. There are responses to responses. Animal awareness is not only a straightforward accumulation of eyes, ears, touch, and other senses but an incalculable synesthesia of mixed senses whose wholeness can be but gleaned by the human consciousness which forms only a part. [PLATE 51]

Both the French paleontologist-priest Pierre Teilhard de Chardin and the Russian atheist Vladimir Vernadsky agreed that Earth is developing a global mind. This layer of thought in the shape of a sphere they called the noosphere, from Greek noos, mind. The aggregate net of throbbing life, from flashing fireflies to human e-mail, is the developing planetary mind. Perhaps, like the brain of a human babe with many synaptic connections that diminish over time, the noosphere is still in its infancy. Polymorphous, paranoiac, confused, yet intensely imaginative, the thinking layer of Earth that is largely the unexpected product of animal consciousness, may now be in its most impressionable stage.

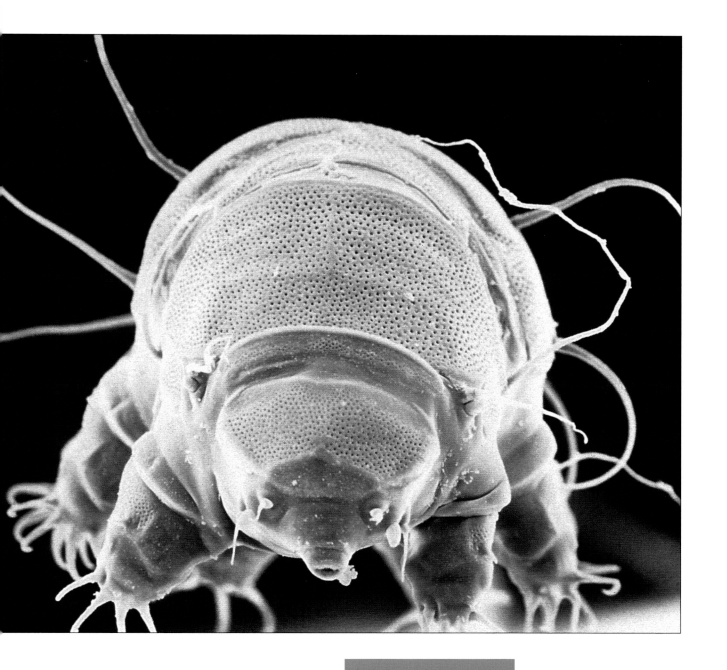

So, what is life?

Life is evolutionary exuberance; it is what happens when expanding populations of sensing, active organisms knock up against each other and work things out. Life is animals at play. It is a marvel of inventions for cooling and warming, collecting and dispersing, eating and evading, wooing and deceiving. Life is awareness and responsiveness; it is consciousness and even self-consciousness. Life, historical contingency and wily curiosity, is the flapping fin and soaring wing of animal ingenuity, the avant-garde of the connected biosphere epitomized by members of Kingdom Animalia. ■

7

Flesh of the Earth

I hold that the fruit of the Tree of the Knowledge of Good and Evil was Soma, was the *kakulj,* was *Amanita muscaria,* was the Nameless Mushroom of the English-speaking people. The Tree was probably a conifer, in Mesopotamia. The serpent, being underground, was the faithful attendant on the fruit.

— R. Gordon Wasson

Truffles…are to be called certainly nothing other than flesh of the earth. Best at spring and more often during thunder, they are said to arouse dying love.

— Franciscus Marius Grapaldus

The Underworld

Academics still often partition life into zoology, the study of animals, and botany, the study of plants. But what of pink molds, single-celled yeasts, puffballs, morels, and psychedelic mushrooms?

Fungi have been lumped with plants because they aren't animals. Medieval scholars working in a three-kingdom system suggested they were zombielike, half-dead forms straddling the mineral and plant kingdoms. Until quite recently the scientific term for fungi has been Mycophyta—from Greek *mykes* (fungus, akin to mucus) and *phyton* (plant). Although none photosynthesize, like plants, some fungi are rooted. But they are best classified as unique in their own, solely fungal kingdom: the Mychota. "And fungi were fungi," wrote the Japanese poet Jun Takami, "they're like nobody else on Earth."[1]

In the English-speaking world the prototypical fungus is a dark, dank toadstool, dimly associated with witches, smelly feet, and refrigerators, and generally to be avoided. "Fungi," declared the eighteenth century French botanist S. Veillard, "are a cursed tribe, an invention of the devil, devised by him to disturb the rest of nature created by God."[2]

Fungi do require sex to form the morel or the mushroom, but they can reproduce without it. Because they don't photosynthesize, they can live in utter darkness. Their vampiristic existence often requires them to do so—sometimes on rather scarce resources of food and water. Reversing the animal technique of taking in food and then digesting it, fungi digest food outside their bodies. They then absorb the nutritious particles through their membranes.

Fungi differ from all other life. [PLATE 52, *page 142*] Unlike plants and animals they form no embryos. They grow from tiny propagules, packages called spores. Upon moistening, the spores form threads, thin tubes, the hyphae. Yeast cells (used in brewing beer and raising bread) bud off single cells. Lacking the whiplike structures of undulipodia, neither single-celled nor multicellular fungi ever swim. Some, called by the fancy name laboulbenomycetes, indulge in fungal sex to form spores that disperse on insect legs. Spores of others attach to mammal fur, are sneezed out or drift in wind. When hitchhiking spores come to rest and sense moisture, hyphae begin to grow up, down, and sideways. Like plants and animals, fungi are made of nucleated cells. Like plants (but not animals) they possess tough cell walls. Fungus cell walls are made of chitin, a nitrogen-rich carbohydrate; plant cell walls are made of cellulose. Many fungi have passages in their cell walls that allow mitochondria, nuclei, and other organelles to move between cells. Some lack cross walls altogether and are more a growing mass of tubes than multicellular individuals.

Fungi break down dead and sometimes live bodies. For more than 400 million years they have been settling and growing on a huge variety of foodstuffs other organisms eschew. A few grow in the sea or underwater, but they are basically land lubbers. Fungi were among the first organisms to make use of terrestrial environments, enabling the development of many other land dwellers. That fungi prefer land was proved when scientists returned to *Alvin*, a sphere-shaped submersible vehicle whose mission to the bottom of the ocean ended when its lifeline to the mother ship snapped. Two years later an intact, if soggy, sandwich was discovered on board the sunken scientific vessel. In contrast, even inside a lunch box any sandwich would be devoured if it were abandoned for two years on a picnic table at Earth's surface. Fungal spores are found virtually everywhere in the air but do not grow on the sea bottom, perhaps owing to an overabundance of saltwater.

Converting waste and corpses into resources, making nutrients available to land life, fungi are invaluable to global metabolism. Fungi thrive amid nature's informal coffins as undifferentiated masses; each fungal thread is bounded, but the expanding organism as a whole—a system of tubes—has no clear borders. Fungi, with their threads penetrating food today but excised by environmental contingency tomorrow, are truly fractal organisms.

The fungal feature commonly called toadstool is really only the minuscule tip of an amorphous underground web of living threads ("hyphae"). A mold growing on bread or fruit demonstrates the typical fungus's lack of regular boundaries. The fungal thallus, an old word for plant tissue undifferentiated into root, stem, or leaf, is also called a mycelium or mycelial network. As the verb "dial" has come to mean "depress buttons on a touch-tone phone," so botanical terms—fruiting body, spore—applied to fungi persist despite their inappropriateness.

Lacking discrete borders, fungi sprawl. Consider the tree-root colonizer *Armillaria bulbosa*. This organic being locates fresh food by continuous exploratory growth. A fungal clone, able to recognize itself from other expanding subsurface fungi, *Armillaria* has territorialized the forest underworld with such inexorable fortitude that, after an estimated 1,500 years of growth, one sample growing beneath virgin coniferous forest in Crystal Falls, Michigan, now encompasses some thirty-seven acres. This individual fungus has an estimated weight of more than eleven tons. A single individual, its genes have been sampled and compared throughout and determined to be the same. Such genetic stability is impressive. Fire, forest succession, and changes in food availability, have isolated parts of the sprawling *Armillaria*, which, nonetheless, maintains its genetic integrity. [PLATE 53, page 143] Are the broken-off pieces separate organisms? Or should they be regarded as the dispersed limbs of a single subterranean being? Reminiscent of a Stephen King novel, the great chunks of biomass

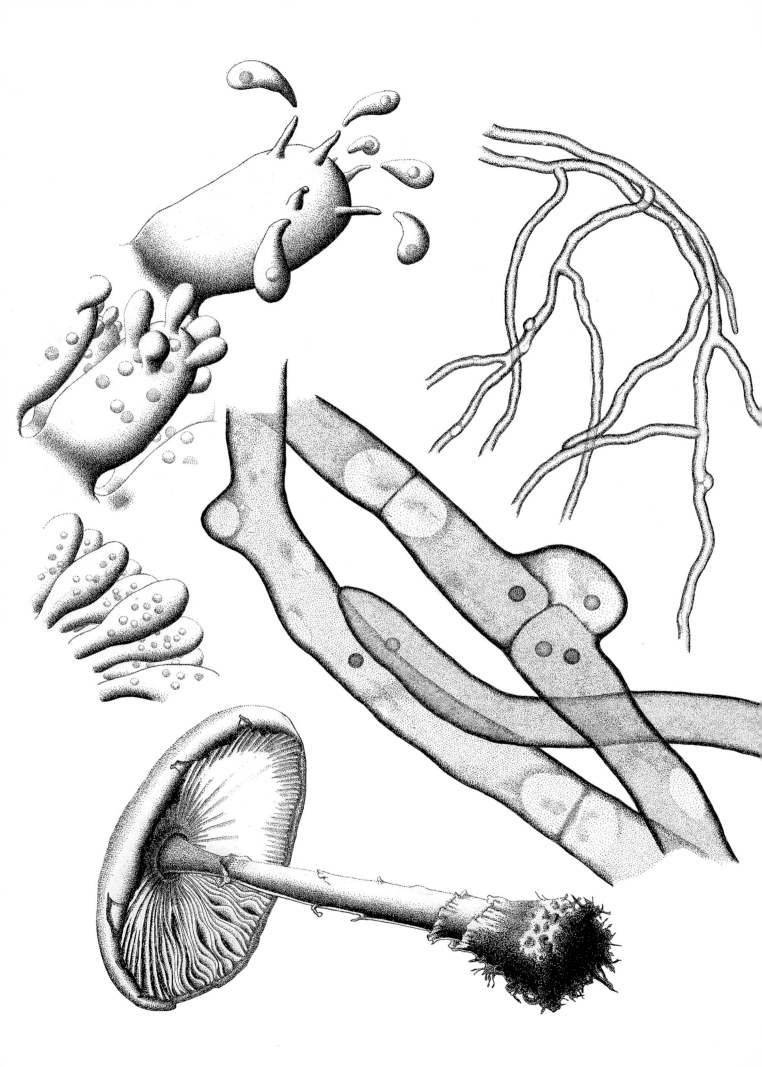

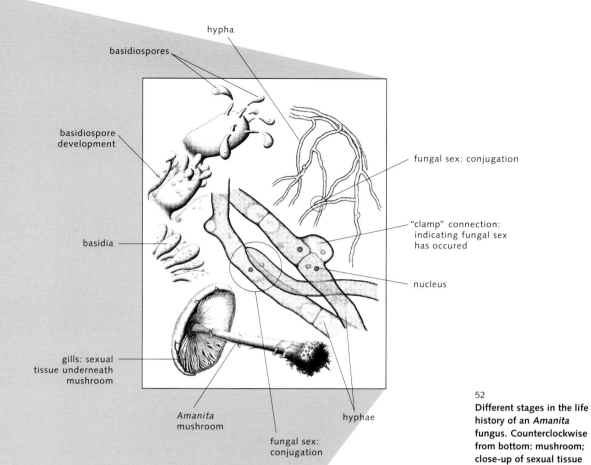

hypha

basidiospores

basidiospore
development

basidia

gills: sexual
tissue underneath
mushroom

Amanita
mushroom

fungal sex:
conjugation

fungal sex: conjugation

"clamp" connection:
indicating fungal sex
has occured

nucleus

hyphae

52
Different stages in the life history of an *Amanita* fungus. Counterclockwise from bottom: mushroom; close-up of sexual tissue or basidia found in the gills; basidia giving rise to drop-like basidiospores; hyphae that grow from spores; nuclei passing through the hyphae in the fungal sex act known as conjugation.

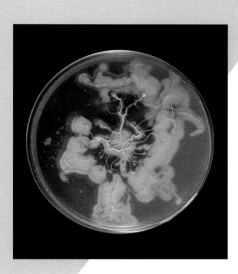

53
***Armillaria gallica* on malt extract agar in a petri plate. The fungal root-like hyphal threads (rhizomorphs) can be detected in a tangle in the center.**

continue to thrive, unperturbed by their multiple amputations.

"Although fungi have acquired notoriety as agents of disease and as producers of psychedelic toadstools," writes scientist Clive Brasier, "their vegetative structure, the mycelium or fungal thallus, has a somewhat lower public profile."[3] The low profile is literal, as most of the mycelium exists in an extensive network, out of view beneath the soil. Great mycelia of foraging hyphae thrive beneath the forest trees. The living threads called hyphae tend to fuse. After "having sex," they eventually form mushrooms or mold tissues that in turn undergo meiosis, forming spores. These disperse over forest and field to grow again in search of mates.

Each mycelial network is a fungal clone, the far-reaching offspring of a single genetic line. Above ground, fungi produce air-borne spores, some of which you no doubt are inhaling now. When they land, the spores grow wherever they can. Sprouting networks of tubes, hyphae, into moist substratum, the fungi once again produce copious quantities of disseminating spores, spreading their strange flesh through the soil they help create.

Kissing Molds and Destroying Angels

No one knows how many species of fungi exist. Some say a hundred thousand. Others estimate one and a half million. Mycologist Bryce Kendrick of Waterloo University in Canada claims that today fungi are more diverse than plants but less so than animals.

As with the other four kingdoms—Monera (bacteria), Protoctista, Animalia, and Plantae— Mychota (fungi) have been arranged into phyla. There are five major groups or phyla of fungi. The zygomycotes (from Greek *zygon*, "twin" or "pair") or mating molds lack cross walls separating their cells. Mitochondria and nuclei travel easily through their open hyphal tubes. In zygomycote mating, special hyphae—the gametangia—grow toward one

another and fuse. From this fusing of gametangia comes the production of resistant spores. Once the hyphal ends lock, nuclei flow through the tubes and join, probably in pairs. When the mating is over, meiosis occurs, producing darkening spores in the head of a black spore holder. *Rhizopus stolonifer*, the most common of the black bread molds, is one example of a zygofungus.

Most molds (such as the pink bread mold *Neurospora*, *Claviceps*, and ergot fungi) and most yeasts (such as *Rhodotorula* and *Saccharomyces*) are ascomycotes. They form asci, sacs or capsules that develop when hyphae of compatible genders "kiss" and permanently fuse. The complex tissues and sexual spores produced from such liaisons is "mold." The hyphal threads, the nonsexual fungal body parts, are there but invisible to the naked human eye.

Ascofungi eat and degrade resistant plant and animal compounds, such as the cellulose and lignin of wood, the keratin of finger nails, and the collagen in mammal bones and connective tissue. By breaking down such compounds, these fungi release carbon dioxide, ammonia, nitrogen, and phosphorus to the rest of the biosphere. The evolution of wood placed great selection pressure on land-dwelling fungi to invent ways to degrade lignin, thus ensuring a coevolved biospheric cycling of matter. Some scientists posit that a lag in fungal evolution contributed to the worldwide accumulation of coal in the late Paleozoic era.

Humans everywhere are most familiar with members of the third fungal phylum. [PLATE 54] The basidiomycotes have reproductive structures called basidia, which resemble clubs (basidion is Greek for "club"). Familiar gilled mushrooms bear these spore-releasing basidia on their lower surfaces. Basidiomycotes include everything from the common supermarket mushroom (*Agaricus*) to the *Amanita virosa* (destroying angel) and its cousin, *Amanita caesarea* (a favorite of the Roman emperor Claudius). Among their numbers are to be found

54

Russula paludosa. Phylum: Basidiomycota. Kingdom: Fungi. This relatively common forest mushroom is connected to the roots of nearby trees with which it lives symbiotically.

55

Schizophyllum commune. Phylum: Basidiomycota. Kingdom: Fungi. The basidia of this mushroom are borne on the white double lines of the gills shown in the photograph.

giant puffballs (up to two feet in diameter), earth stars (looking like tiny breasts decorated with the leaves of a jester's cap), smuts, rusts, and jelly fungi [PLATE 55].

A fourth phylum, the deuteromycotes, consists of molds that form neither basidia nor asci. The deuteromycotes probably lost this ability when they lost sex. But they are nevertheless reproductive wizards, capable of the ceaseless production of airborne propagules. These organisms, also called conidial fungi or Fungi Imperfecti, reproduce by conidia, thin-walled cells that break off from the tips of ordinary hyphae. Others lack special reproductive structures; any part of the body, any hyphal thread or mycelial mass may break off to reproduce.

The fifth phylum of fungi consists of low-lying photosynthesizers—the lichens. Lichens are one of the most striking examples of symbiosis. They are also among the most successful fungi. Like the bacterial merger that led to algae, the lichen is a combination of fungus and alga (or sometimes, cyanobacterium). The result: an altogether new life form that takes advantage of the alga's ability to make its own food and the fungus's ability to store water and fend off the elements.

56

Cladonia cristatella, a lichen. A permanent merger between members of the fungus kingdom and photosynthetic members of the Protoctist kingdom, in this case the green alga *Trebouxia.* Thousands of different fungal species combine with a few photosynthetic partners to produce a great variety of lichens, such as this red-tipped British soldier lichen.

57

Isolated fungal partner of *Cladonia cristatella,* shown in the preceding Plate 56. The synergy of fungus and alga produces a structure that could not have been projected on the basis of simple addition.

58

Trebouxia, an alga. Phylum: Chlorophyta. Kingdom: Protoctista. The photosynthetic component of *Cladonia cristatella* seen in Plate 56.

Cross-Kingdom Alliances

All lichens—some 25,000 different kinds is an estimate—result from cross-kingdom couplings between fungus and either green alga or blue-green bacterium. Many lichens even harbor both types of photosynthesizers at once. Dwelling on bark and clinging to gravestones, sheer cliffs, and other sunlit places unavailable to less enterprising organisms, lichens have created a cozy niche for themselves. As they grow, they slowly turn solid rock inside out, into crumbling soil and living earth.

Divided, the fungal gray and photosynthetic green components of a lichen look nothing like one another. Nor does either member resemble their extraordinary composite. [PLATES 56, 57, and 58] The result of symbiosis, far from being predictable by simple addition, is a noncumulative surprise.

With so many different lichens, each representing a permanent tryst between fungus and photosynthetic life form, the phrase "long-term relationship"

takes on new meaning. When the photosynthetic partner of a lichen is removed, sugar excretion by this pigmented partner stops and does not begin again even if extracts of the sugar-inducing fungus are reintroduced. Somehow, algae and fungi sense each other's whole-body presence as they form an enterprising, complex partnership that depends on the history of the relationship. Like animal cells, the algal and fungal cells in a lichen communicate metabolically. Unlike most animals, however, the size and shape of any variety of lichen is not precisely fixed and the extent of their tissue complexity is limited to one or a few tissue layers. Lichens, however, surpass animals in their longevity; individual lichens may be 4,000 years old.

The ultimate interkingdom alliance, although not yet a lichen, may be the one that exists today in Antarctica. Seventy percent of Earth's fresh water exists in Antarctica, but the water there is tied up as

59
Mycorrhiza, a synergistic structure, a symbiotic protuberance produced by fungus and plant, in this case the fungus *Genea hispidula* on roots of the beech tree, *Fagus sylvatica*.

ice, and the relative humidity at these forsaken outposts rarely exceeds thirty percent. The few regions of Antarctica that are ice-free are thus deserts—and they are the driest places on Earth. Yet in the barren expanse of these cold deserts, so-named endolithic fungi grow with green algae inside the rocks. Feeding off the algae with which they loosely reside (and which obtain sunlight through the translucent rock crystals), endolithic fungi derive their water from rare but adequate melting frost.

Life can evolve suddenly, by jumps, when separate parties unite. Interkingdom alliances between fungi and algae produced lichens; a similar alliance may have been crucial to the development of the first forests. Root growths called mycorrhizae result from the dual growth of fungi and plants. [PLATE 59] Mycorrhizae provide the autotrophic plant partner with mineral nutrients, whereas the heterotrophic fungal partner is supplied with photosynthate food.

Galls on the stems of *Solidago,* goldenrod. Galls, "disease" structures, may represent symbiotic organs in an early phase of development. Such bulbous growths occur when plant tissue interacts with fungi, insects, and perhaps bacteria. It has been theorized that galls were the evolutionary predecessors to the first fruits.

Mycorrhizae—rounded, stubby, often colorful roots—are symbiotic, a dynamic structure produced by plant *and* fungus. More than 5,000 distinct mycorrhizal fungi have been discovered. Most of the associated plants seem to depend on these symbioses for their supply of soil phosphorus and nitrogen. Mycorrhizae look neither like plant root hairs nor like the fungus's mycelial network. They are synergistic, emergent structures, crucial to recycling. A single large tree may have a hundred different mycorrhizae, produced by distinct fungi, living in its roots.

Plants and fungi joined forces from the very start of life on land. Some of the oldest plant fossils in the world retain evidence of symbiotic fungi. Lacking leaves, branches, and twigs, the first land plants were little more than upright green stalks. Peter Atsatt, botanist at the University of California at Irvine, and Kris Pirozynski, former mycology researcher at the Museum of Nature in Ottawa, both contend that successful colonization of the land by the ancestors of modern land plants would have been impossible without root fungi. Today fungi are still synergistically intertwined in the roots of more than ninety-five percent of plant species. The algal ancestors of plants may not have been able to come ashore without nutrient-procuring fungi. The primeval arboreal carpet, the first forest floors, appear to have been created not just by plants but by plants and fungi acting together.

Kingdom Plantae is (and always has been) almost entirely terrestrial. The algal ancestors of this kingdom, of course, emerged from aqueous environments, but most of the descendants kept to the land. Early plants on land had to overcome almost insurmountable odds. Land then was mercilessly irradiated and depleted in the forms of nitrogen and phosphorus salts absolutely required for plant growth. Moreover, the land was and is an unreliable provider of that vital resource, water. Until the Silurian, when plants and fungi began to occupy

the land, cyano- and many other kinds of bacteria had the desolate continents to themselves.

Kris Pirozynski has hypothesized that fruits—whose colors, flavors, and aromas still cast an aesthetic spell over our primate brains—evolved by way of interferences from the fungi and animal kingdoms. His hypothesis attempts to explain the gap in the fossil record between the spread of flowering plants and the appearance of fleshy fruits a good forty million years later. Pirozynski envisions the first fruit appeared when fungal genes were transferred into plant chromosomal DNA. This is similar to what occurs in crown gall disease. Galls are symbiotic tissues formed by insect, bacterial, or fungal growth on plants. [PLATE 60] They are bloated, sometimes slightly monstrous-looking tumors found mainly on shrubs and trees—and some look remarkably like fruits.

Crown gall disease is caused in many plants by penetration of *Agrobacterium*. *Agrobacterium*, a kind of bacterium that lives in soil, bears plasmids (short pieces of DNA) that can enter into the cells within roots and stems of susceptible plants, bringing bacterial genes into the plant's nuclei. Biotechnology firms use *Agrobacterium* to introduce desirable genes into crop plants. Pirozynski speculates that fungal genes may have infiltrated plants in similar fashion. That fungal infections were important to the abrupt, relatively late appearance of fruits in many different flowering plants in the Cretaceous fossil record remains an intriguing hypothesis. That galls are examples of plant-fungal synergy, however, is an established fact.

Underbelly of the Biosphere

Fungi resemble animals in that, unable to produce food, they depend for nutrition on the bounty of others. From an ecological viewpoint, however, the two kingdoms differ markedly. Fungi are indispensable to the formation of soil, breaking down intractable rock. They help lay down the carpet of spreading life. They are the underbelly of the biosphere.

Without fungi, plants and eventually all animals would be starved for phosphorus (part of the autopoietic imperative as an essential component of RNA, DNA, and ATP). Fungi mediate the interstices of the food web. The Arab scholars who classified fungi midway between the plant and mineral kingdoms had a point. When fungi take over a body, its material nature is quickly revealed. Bodies become carbon-rich humus. Fungi decompose corpses and feed on living tissue, such as the skin of sweaty feet. For more than 400 million years their spores have been settling and sending mycelial networks through a global smorgasbord of foodstuffs. Recycling the dead, they are the garbage collectors of the biosphere.

Fungi break down bread, fruit, bark, insect exoskeletons, hair, horn, camera lens mounting compound, film, skin, building beams, cotton, feathers, the keratin of finger nails and scalp. Like portable cleanup crews, fungi are transported around the world by airborne spores. Almost nothing is exempt from their gastronomic ministrations. Indeed, their zeal for recycling is so great that many begin even before an organism has died. In maladies such as athlete's foot, jock itch, and ringworm, fungi get ahead of themselves in the job of redistributing elements of the biosphere.

Whether growing on human epidermal cells, breaking down the cellulosic fibers of cloth with cellulase enzymes to create mildew, or making green-gray spores as a penicillium mold begins its colonization of grapefruit, fungi digest substances left over by others. What seems like "decay" to us is, to fungi, the healthy growth of new offspring. Without fungi and bacteria breaking down complex macromolecules, the corpses of plants and animals would pile up, thereby taking phosphorus and nitrogen out of circulation.

Fungi, on land, perform most of the waste management function in the biosphere. Unlike ordinary people, who have survived as polluting nomads for generations, dumping and moving on, the biosphere cannot simply put its refuse out on the planetary curb. On Earth, garbage does not go out, but around. Humans are only now approaching the level of sanitary efficiency mastered 400 million years ago by the fungi: fungi do not simply remove garbage, they recycle it. Supplementing the bacteria, fungi recycle carbon, nitrogen, phosphorus, and the like; on a plant- and animal-dominated continental landscape they expanded planetary autopoiesis to dry land, changing Earth's surface forever.

Hitchhiking Fungi, Counterfeit Flowers, and Aphrodisiacs

As rapid recyclers fungi often send the rest of us mixed messages. Mediating at the interstices where garbage becomes food and corpses become fertilizer, fungi can cross up animal nervous systems.

Ever since amphibians and their descendants ambled onto land, we animals have had to contend with fungi. Indeed, for millions of years fungi and animals have coevolved. Our primate ancestors dwelled in forests and tasted many foods. Some were poisonous, others mind altering. When spores or hyphae, resisting digestion, can pass through animal intestines, it may be advantageous to them to be eaten. Those animals who find fungi delicious often offer them a nonstop free ride to the soil. For their part, animals that detoxified or vomited poisonous fungi survived to unwittingly disperse them.

As language evolved, social prohibitions against ingesting possibly dangerous fungi developed. So did sacred rites involving their use. The attempts of societies to try to rid themselves of drugs deemed threatening is reminiscent of the body's autonomic response to rid itself of certain fungal "foods." But due in part to the ability of certain species to produce mind-altering trips, fungi will never be completely eliminated from the body politic. Fungi are an entrenched part of sentient life in the biosphere.

In their ancient capacity as sanitary engineers fungi have evolved some rather startling relationships with members of other kingdoms. *Phallus* and *Mutinus* are penis-shaped stinkhorns of the order Phallales, whose stench, reminiscent of decaying meat, attracts flies. The flies that alight on them carry sticky fungal spores away on their legs. *Pilobolus crystallinus* is a kind of mating (or "kissing") mold whose favorite habitat is horse dung, rich in undigested cellulose, nitrogen, and other nutrients passed up by animals. Indeed, dung is so valuable to fungi like *Pilobolus* and other mating molds that they have evolved a clever stratagem to get there first: they arrange to be eaten, lying in wait amid blades of edible grass. Mature *Pilobolus* does not linger in the dung. Its sporeheads absorb water from the feces, internally building up intense pressures. Tensed for action, the sporeheads, unbranched structures two to four centimeters in height, aim toward the light. When internal

pressures exceed 700 kilograms/cm² the sporeheads pop out of the manure — and land several meters away, in grazeable grass. Here the long jumpers, an inspiration to the internationally acclaimed Pilobolus dance troop, can be eaten by helpful, munching horses.

Fungi have evolved other ruses. One yellow rust fungus, a basidiomycote, invented the insect-fooling stratagem of mimicking a flower. After infection the rust fungus induces rock cress (a plant in the mustard family that grows in Colorado's mountain meadows) to metamorphose and mimic the common yellow buttercup. Rock cress has unspectacular, droopy flowers but, once fungally infected, it elongates and makes a nectar-rich rosette — a rigid uplifted "flowerhead" that attracts would-be pollinators. Instead of pollen, the visiting insects pick up spores from these counterfeit flowers.

More than forty fungi species, such as *Panellus*, a delicate capped stalk, glow in the dark. Why they luminesce is not known, but a good guess is that in doing so they attract animals such as nematodes, tiny translucent worms. The worms eat the fungus, excreting and spreading the undigestible sticky fungal spores.

As ancient pioneers of land, fungi work with more recent settlers to propagate themselves and their offspring. The extraordinary willingness of fungi to perpetuate themselves by any means, and through any ends, possibly climaxes in the evolution of a particular fungus-and-ant agricultural system. Ever loyal to the fungi they cultivate for food, ants of the genus *Atta* evolved wheelbarrow-like depressions in their backs for carrying spores, which they then fertilize and feed with chomped-off bits of leaves, bark, and other plant material. [PLATE 61] These insects farm fungal spores as if they were seeds, carefully weeding out debris that might spoil their underground gardens. As with the fungal flower mimics, this cross-kingdom association brings to mind science fiction stories of beings enslaved to

So, what is life?

Life is a network of cross-kingdom alliances, of which Kingdom Mychota is a willing and crafty participant. Life is an orgy of attractions, from the trickery of counterfeit "flowers" to the strange allures of truffle and difficult-to-swallow hallucinogens. As fungi, life seeks out the underworld of soil and rot no less than the sunny vistas overwhelmed by photosynthetic beings. Life is self-renewing and fungi, as recyclers, help keep the entire planetary surface brimming with life. Transmigrating matter, molds and mycelia have found their calling. Creating and destroying, attracting and repelling, undertaking and overturning, they are part and parcel of *terra firma*. ■

8
The Transmutation of Sunlight

Tyger! Tyger! burning bright
In the forests of the night,
What immortal hand or eye
Could frame thy fearful symmetry?

— **William Blake**

If one has the patience, and the courage, to read my book, one will see that it contains studies conducted according to the rules of a reason that does not relent . . . but one will also find in it this affirmation: *that the sexual act is in time what the tiger is in space.* The comparison follows from considerations of energy economy that leave no room for poetic fantasy, but it requires thinking on a level with a play of forces that runs counter to ordinary calculations, a play of forces based on the laws that govern us. In short, the perspectives where such truths appear are those in which more general propositions reveal their meaning, propositions according to which *it is not necessity but its contrary, "luxury," that presents living matter and humankind with their fundamental problems* . . . freedom of mind . . . issues from the global resources of life, a freedom for which, instantly everything is resolved, *everything is rich.*

— **Georges Bataille**

The rays of the Sun determine the chief characteristics of the mechanism of the biosphere. They have completely transformed the face of the Earth. To a great extent, the biosphere is a manifestation of them. It is a planetary mechanism which converts the radiations into new and varied forms of energy, which completely changes the history and destiny of our planet.

— **Vladimir Vernadsky**

Green Fire

The ultimate source for all life's energy, growth, and behavior is the sun. Burning like a cool green fire, photosynthetic beings transmute sunlight into themselves. [PLATE 65] Protoctists (coccolithophorids, diatoms, seaweeds) are the main transmutors in the sea; plants the main ones on land.

Plants represent a high point in bacterial coevolution. They have raised the biosphere to a higher dimension — up to a hundred meters from the soil surface. Yet they are newcomers to the photosynthetic guild. Plants have only dwelt on Earth for the last 450 million years. Evolving from algae, plants — almost exclusively land beings — went on to green the continents.

The blue whale, 26 meters long and weighing 180,000 kilograms, is the most massive animal ever to have lived, heavier by far than the largest dinosaurs. Nonetheless, next to behemoths of the plant world — such as the giant sequoia, who may weigh 2,000,000 kilograms — even whales are light. One clone of the quaking aspen, *Populus tremuloides*, is estimated to contain 47,000 trunks. Nominated by University of Colorado biologist Jeffry Milton as perhaps the biggest individual organism on the planet, this dispersed but connected tree covers 43 hectares in Utah. It is estimated to weigh 6,000,000 kilograms. [PLATE 66]

Books come from plants. So do boardwalks, oak desks, hashish, cotton shirts, chewing gum, coal, myrrh, clapboard houses, chocolate. Plants are the source of morphine, codeine, heroin, and other drugs similar to endorphins, pleasure-giving chemicals produced naturally in the mammalian body. Bark of *Salix*, the willow clan, gives us salicylic acid, aspirin; other plants make not only analgesics but astringents, antifungals, antispasmodics, pigments, caustics, cardiovascular agents, expectorants, diuretics, fumigants, hemostatics, insect repellents and toxins, perfumes, and anti-asthmatics.

Plants have been such a deeply embedded part of

65
A chloroplast, the intracellular structure which carries on photosynthesis. This close up, magnified 45,000x, was taken with an electron microscope. A hallmark of plant life, chloroplasts, as has recently been "proven" by genetic comparison, evolved from cyanobacteria that greened the world long before the origin of plant life proper. Algae and plants all seem to have evolved after larger cells merged with smaller ones. The large cells fed on but ultimately failed to digest the free-living cyanobacteria.

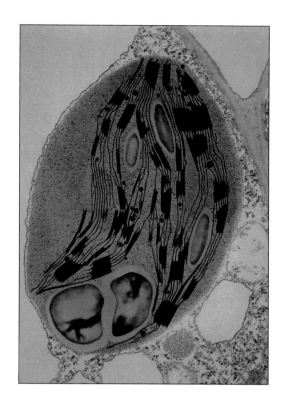

66
Populus tremuloides, quaking aspen. Phylum: Angiospermophyta. Kingdom: Plantae. A stand of quaking aspen in the San Juan Mountains, Colorado. A similar stand in Utah has been nominated as the largest "organism" on Earth. It has been so nominated by considering each genetically identical tree, the stem of a many-treed body. The aspens in the picture, turning color simultaneously, are not as extensive as the 43-hectare, 6,000,000-kilogram stand sharing a single root system in Utah.

the human environment that we now hardly notice them. Unless a bouquet of long-stem roses or box of chocolates arrives on the doorstep, plants are taken for granted. Even then, meaning or mood is evoked by the plant product as symbol rather than the plant itself.

Plant life presents us with an extraordinary richness of sights, smells, and tastes. The seasonal bursts of fragrant flowers have a beneficial psychological effect for dwellers outside the tropics; the mere sight of hills of undulating grass can produce serenity.

Of the nine recognized phyla of plants, only one has flowers. But that one phylum is so diverse that it is thought to account for more than half of all plant species. A full documentation of all species in the three hundred families of flowering plants would be such a monumental task that it has never been undertaken. "Such a listing," writes botanist Frits Went, "would have to describe about a quarter million known plants; to compile it, all the taxonomic botanists in the world would have to work together for years and years, and the finished product would have perhaps half a million pages, enough to cover a whole wall in a library."[1]

Plants were not, however, the first form of "green fire" on land. Fifty miles southwest of Las Vegas an 800-million-year-old fossil soil is preserved in rock. The carbon content indicates some kind of ancient photosynthetic life. From another spot in the American Southwest, eighty miles northeast of Phoenix, Arizona, even older samples of fossil soils were collected, corroborating a hypothesis by Susan Campbell and Stjepko Golubic of Boston University that photosynthesis on land began in cyanobacterial form 1,200 million years ago or earlier. Paleontologist Robert Horodyski of Tulane University in New

Orleans and geologist L. Paul Knauth at Arizona State University in Tempe contend that land was abundantly covered by photosynthetic microbes in the late Proterozoic eon.[2]

Not until the Silurian appearance of true plants, with their alternation of spore propagules (formed by meiosis) and gametes in sex organs (that fuse to make embryos), was there a full-fledged escape, probably at first seasonal, from the algal necessity of dwelling in water. [PLATE 67] Freeing itself of water, life on land evolved internal means of support and grew up on land. Land plants made a water-pressure system of structural support from the cellulose molecules found in bacteria and algae. Later, they evolved a stronger substance that, when combined with cellulose, would remain elastic but strong and supportive, even in dry conditions. This substance, lignin, is the chemically complex polyphenolic that gives woody plants their woodiness. With lignin, the biosphere began its vertical climb, extending life's realm to another, third dimension over the land. Biologist Jennifer Robinson has suggested that the great piles of coal left in the Earth's crust owe their existence to the lag after plants had invented lignin but before fungi had evolved means of decomposing it.

Just like other life forms, plants came from microbial predecessors. Their heritage is of photosynthesis, but they need not be photosynthetic. Some plants, even those with leaves and fruit, have abandoned the way of green fire, no longer photosynthesizing. Like those eyeless, subterranean mole rats that no longer see because they do not need to, some white plants have outgrown their dependence on direct sunlight. For example, *Epifagus* (beech drops) and *Monotropa* (Indian pipes) [PLATE 68, *page 162*] eke

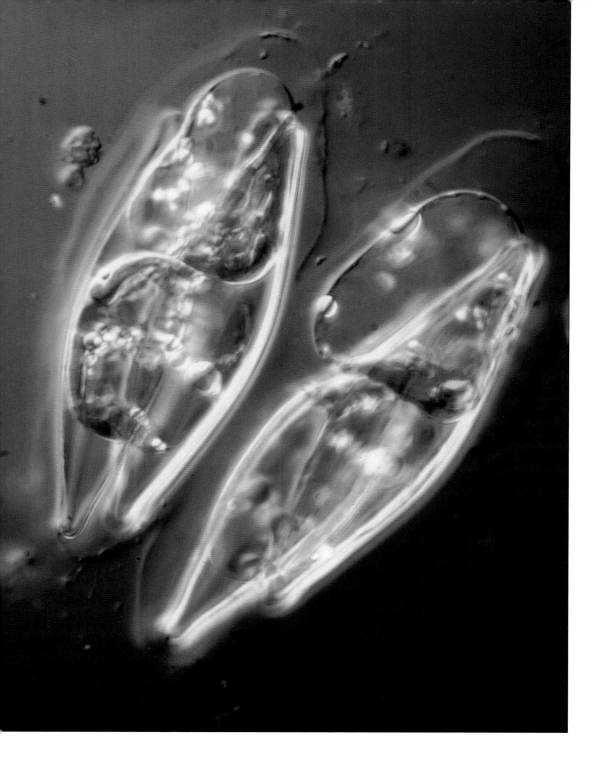

67
Navicula cuspidata, a diatom alga. Phylum: Bacillariophyta. Kingdom: Protoctista. A diatom undergoing meiosis and gamete formation. Algae, like all protoctists, dwell in water. Plants eventually escaped water to venture onto land by evolving waterproofing such as waxy cuticles and structural support such as lignin.

68

Monotropa, Indian pipes. Phylum: Angiospermophyta. Kingdom: Plantae. All plants develop from embryos but not all plants are photosynthetic. The Indian pipes in this picture are albino parasites that have lost their chlorophyll but have retained ghostly remnants in the form of blanched plastids.

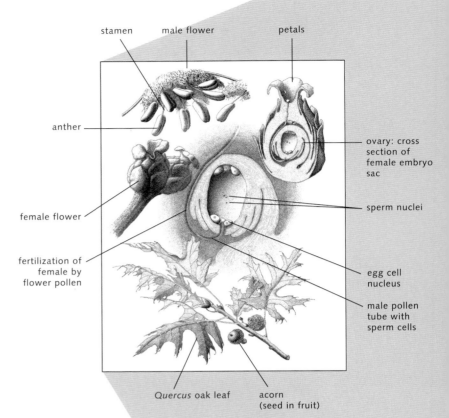

stamen — male flower — petals

anther

female flower

fertilization of female by flower pollen

ovary: cross section of female embryo sac

sperm nuclei

egg cell nucleus

male pollen tube with sperm cells

Quercus oak leaf — acorn (seed in fruit)

out nutrients by subvisible fungal threads to their benefactors, neighboring green forest trees.

The distinguishing feature of plants is not photosynthesis per se but that they all grow from spores at one stage in their life cycle and from embryos in another. The plant embryo, found deep in the maternal tissue, is the diploid product of sexual union. While it is, like an animal embryo, formed from the fusion of a male propagule with an egg held in a female organ (the "archegonium" in plants), neither the egg nor the sperm of a plant is produced by meiosis. Plant embryos form when male pollen tubes or swimming sperm penetrate tiny female plants or the female parts of hermaphroditic plants. These tiny female plants are haploid and grow from haploid spores buried in the mother's haploid tissue. The mature plant that grows from the embryo does not make gametes (as mature animals do); rather, it makes, through meiosis in its diploid body, spores. The spores grow into either male or female haploid plantlets that make gametes without meiosis. [PLATE 69]

Plants are sexual beings. Sexual coupling is the act, and the embryo is the structure, that distinguishes them from algae and members of other phyla (lichens, for example) that have sometimes

69

Different stages in the sexual life history of an oak tree, *Quercus*. Counterclockwise from top: oak leaves with two mature fruits that contain the seeds (acorns). The flower shown at center right is magnified and its walls (the ovary) are cut away to reveal the eight nuclei of the double fertilization of angiosperms. The pollen tube has penetrated the embryo sac and let loose three small male nuclei. One will fertilize

the egg nucleus (bottom center) and one will fertilize two of the larger female nuclei to form triploid (3-chromosome set) tissue which will nourish the embryo—hence "double fertilization." The doublet anthers that produce the pollen are shown on their stalks at the lower right. At the lower left a germinated pollen grain that has formed a pollen tube is seen on its way into the ovary of a bisected oak flower.

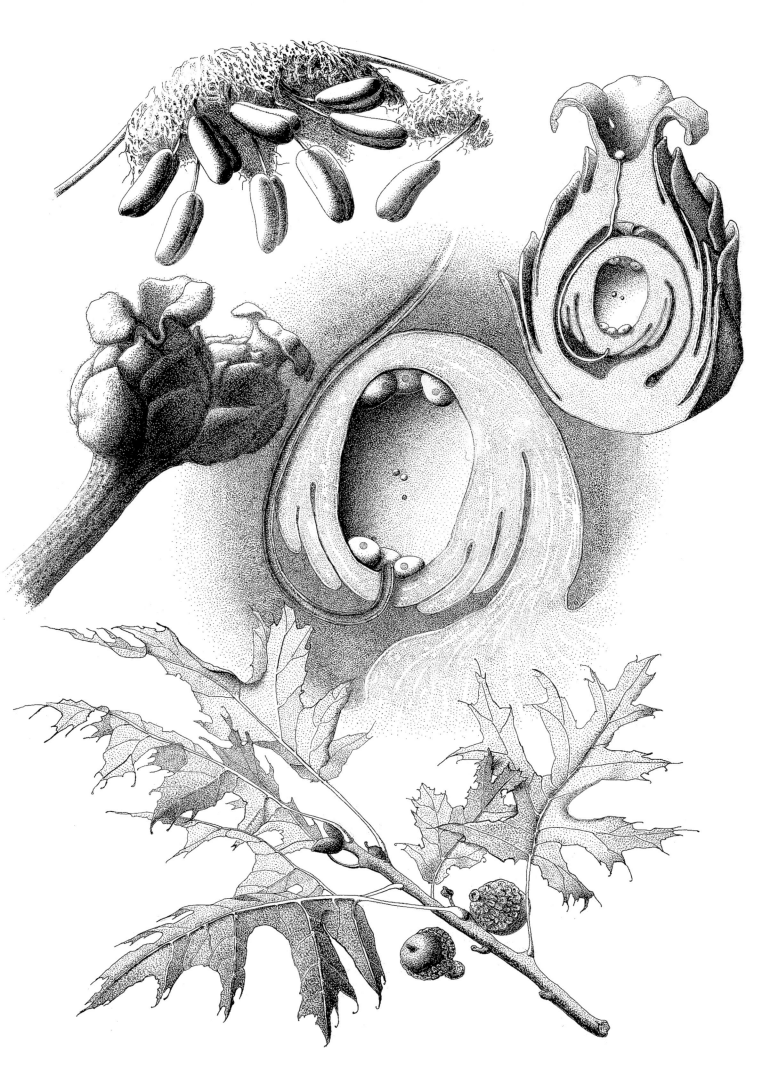

been misrepresented as "plants." Plant sex, however, differs from that of animals: although fertilization of plant egg and sperm nuclei makes embryos, plant meiosis does not make eggs and sperm. Meiosis makes spores. Spores grow into plantlets, each with only a single set of chromosomes. These plantlets, called gametophytes, can grow—unlike an unfused animal egg or sperm. The gametophyte grows by mitotic division of its cells, which carry only a single set of chromosomes.

In cone-bearing and flowering plants the gametophyte (either male or female) is just a tiny structure that is not free-living. The gametophyte forms and spends its life entirely within a cone or flower of the plant that meiotically produced it. Plantlets make their sex organs and their gametes by mitotic cell division. Because they begin with only one set of chromosomes, these cells do not have to change their chromosome numbers to produce sperm or eggs. When the mating cells fuse, doubleness is reestablished and the cycle begins again.

Evolutionarily speaking, however, the failure of the gametophyte to "leave home," so to speak, is a recent condition. Plants of an older lineage, ferns for example, cycle through an alternation of generations in which the small bodies of those with a single set of chromosomes and the large ones with two sets are physically unconnected and strikingly different in form.

Overall, as sexual embryo-forms, plants and animals are more alike than either are like the three other kingdoms (bacteria, protoctists, and fungi). Animals are, however, diploids with a single-cell haploid phase, whereas plants have a multicellular haploid phase—and, unlike the cells of animals, all plant cells contain remnant cyanobacteria.

The Accursed Share

In the enigmatic epigraph to this chapter, Georges Bataille links the tiger with that initiating point of the mammalian life history, sexual intercourse.

He assures us, moreover, that his comparison is rational. It is. The entire unfolding of evolution is a response to an unexportable excess, a growing surplus of sun-derived energy. Both the sex act and the tiger are complexities of the biosphere.

While coitus is a behavior and the tiger a being, together they represent two fates of plants' prodigious reserves. The tiger is poised atop a pyramid of global nutrition whose base is the sun. Even at rest, the tiger represents life's nutritional edge, its carnivorous limit. The tiger, "burning bright" in Blake's memorable phrase, represents the funneling of solar radiation into a highly specific and potentially terrifying form. Coitus employs sun- and plant-derived wealth as animals expend energy to make more of themselves. Bataille further argues that classical economics is mistaken: the general economy is not human but solar. Sun-produced food, fiber, coal, and oil, carbon- and energy-rich reserves are the living foundation not only for bustling animal life but for industry, technology, and the wealth of nations.

The economy comes from photosynthetic life and the sun. Photosynthesizers use solar radiation to produce the cold hard cash of the biosphere. Heat is dissipated, degraded energy lost to space as primordial wealth accumulates. Colorful photosynthetic bacteria, protoctists, and plants the world over produce and "save." Eating them, consumers may "spend" through metabolic activities gathered photosynthetic energy or anabolically (and temporarily) store it in their herbivorous or predatory tissues. Primordial wealth may also end up in long-term storage (or be lost outright) when consumers die and are buried without decay.

Spending has always been a critical problem for life. Greed comes easily within a biosphere whose constituency triumphs as a function of ability to amass the wealth of photosynthesis. Bataille's tiger mercilessly hunts the leaf-eating deer. North Americans now fell trees to print paper money with colored fibers—or submit such bills in return for the

70

Forest floor in Coos County, New Hampshire. Fungi are here externally digesting leaf litter, animal corpses and other debris whose temporarily trapped, ultimately solar-derived energy will thereby be made available to other organisms in the biospere. In this way bacteria and fungi recycle and transport valuable compounds made of carbon, nitrogen, phosphorus, oxygen, and sulfur, spreading the wealth of the solar economy.

striped pelt of that endangered mammal. Photosynthesis creates excess, surplus, a reserve of matter and energy whose uses are as numberless as life is creative.

Bataille perceived that the character of a particular society is determined less by its needs than by its excesses. Wealth creates freedom in both biological and cultural realms. A nostalgia for old Europe, a respect for native American restraint, an admiration for the opulence of Egypt—these are sentiments based implicitly on the recognition that a culture is determined by how its members choose to spend or accumulate its excess. Rome makes its coliseum and basilicas, America its MacDonald's and Disneyland, Egypt its sphinx-guarded pyramids.

In the United States politicians grapple with tax collection, deficit and debt reduction, and public spending. The government prints money that banks lend without having or touching. Stocks, bonds, certificates of deposit, precious metals, and other instruments of finance are owned by investors. But what does it mean "to own"? Humanity does not own what it spends; ownership rests with the biosphere. Checks, credit cards, paper money, and stock certificates are all symbols of a wealth whose source lies beyond technological humanity's means of production. The monetary economy attempts to arrest the solar flux of Earth's economy. Money symbolizes the conversion of photosynthesis, life's energy, into something else—something that can be controlled, manipulated, and hoarded by humans. Perhaps it is no coincidence that in the United States money is green.

The fact remains that without plants the vast majority of animals would starve. Indeed, even with luxuriant plant growth humans and all other animals are destined to die. The grave is a great leveler, and a good reminder that we are owned by what we own. All of us from street sweeper to billionaire pay our dues. The elements of our bodies return to the biosphere whence they came. In the restricted economy of human arrogance and fantasy, individuals may amass great wealth and power. But in the solar economy of biological reality each and every one of us is traded away to make room for the next generation. On loan, the carbon, hydrogen, and nitrogen of our bodies must be returned to the biospheric bank.

A biosphere differs from an organism in that it is essentially closed to influx and egress of materials. Although supplies of carbon usable by life arrived with meteors and comets that penetrated Earth's atmosphere, especially before life took hold, this external source of material today is insignificant. Unlike an organism, eating and excreting, the biosphere has become self-contained. Its materials are limited, used over not up. The luxurious surplus of edible and usable compounds produced by photosynthesis leads to scavengers and predators, organisms killing and eating or cleaning up to survive and grow. The limited material reserves of the biosphere constrain the amount of solar rays that can be transformed into green life. [PLATE 70]

Overall, photosynthetic activity creates a surplus of energy-rich matter that can be hoarded, eaten for growth, or outright squandered. The great planetary

THE TRANSMUTATION OF SUNLIGHT

riches are there for the taking, replenished by the lively conversion of solar energy. It is an understandable but impossible wish to preserve the planet in its "original" state. The pristine nature to which some wish to return is not eternal but rather the green world that supported our ancestors so beautifully that they overpopulated it. Moreover, human spoilage of the lush environments that nurtured us is not evidence of any singular ability to imperil all life on Earth. No single species in the past has ever threatened all the others. Any tendency of one kind to overgrow and despoil was kept in check by all the rest. The essence of "natural selection" is that unstoppable tendencies of one population to grow to the point of environmental degradation will be halted by the growth of others. Human population expansion plays by the same rules: the degraded environment breeds morbidity, high mortality, and ultimately even extinction.

Our evolution has unearthed hoarded organic treasures, such as coal and oil to power cars and heat homes. Wealth in the biosphere ultimately comes from the sun. Organisms die, populations decline, and species become extinct. But the biosphere gets richer. Human burning of fossil fuels, for example, is exploited by plant life. Plants incorporate carbon dioxide released from this burning into their bodies. This is not to say that the current industrial mode of human living may not be dangerous or lead to increases in global temperature. Rather, the conversion to waste of a surplus by one life form has biospheric precedents: far from impoverishing the planet, the waste of one may, in fact, create more wealth for another.

In the strange solar economy individuals die, returning their bodily wares to biospheric circulation. Chemicals used in bodies are not lost. All organisms confront the combined difficulty and temptation of making use of that persistent photosynthetically-derived excess to which Bataille gave the name "the accursed share."

Ancient Roots

The first plants were probably like today's bryophytes. Mosses, liverworts, hornworts, and their kin lack the vertical stature of other phyla, owing to the absence of any system for fluid transport and hydrostatic support. Little more than masses of green cells favoring moist surfaces, bryophytes then and now lack leaves, roots, and seeds.

At the end of the Ordovician period the land surfaces were coarse, populated by low-lying cyanobacteria and soil algae but no plants. Where water was dependable, in rivers and lakes and along the borders of the sea, cyanobacterial mats became thick. In drier locales a sinuous, tough binding of blackish green soil particles and angular bits of rock covered the land. A modern analog of such terrestrial life prior to plants are the desert crusts of Utah, the Gobi desert, and the fields of Iraq. These crusts consist of cyano- and other bacteria, occasional algae, and fungi—all of which are quick to begin the green fire of photosynthesis (or revert to a quiescent state) when given moisture.

Some modern green algae ("chlorophytes"), especially the chaetophorales, have been proposed as similar to the ancestors of plants. Their chloroplasts contain chlorophylls *a* and *b*—the same pigments found in the chloroplasts of plants. Like the sperm tails of mosses and ferns, two quite different sorts of plants, motile chaetophoralean algal cells bear two undulipodia. These green cells have intercellular connections, plasmodesmata, that resemble the perforations through the cell walls of plants. Animal cells join by strengthening contacts utterly unlike the perforations of plasmodesmata of algae and plants. The details of their mitotic cell divisions and their walls made of typical plant cellulose suggest that certain chlorophytes, such as the modern filamentous green alga *Klebsormidium*, resemble the imagined ancestors to plants.

Today ferns, ranging from fewer than three centimeters to over twenty meters in height, still reproduce by aquatic methods of egg and swimming sperm, which they shed into nearby puddles. Even so-called higher plants, such as the *Ginkgo* (a showy tree with fan-shaped leaves and stinking cherry-like cones, indigenous to steep slopes in eastern China), bear their ancient heritage in the form of undulipodiated sperm. The many tails of a single sperm in cross-section are undulipodia with the same 9(2)+2 symmetry of microtubule arrangement that is found in motility structures from algal swim tails and *Paramecium* cilia to bull sperm and the fine hair cells of the human lung.

The oldest well-preserved plant fossils are from black cherts at a quarry in Rhynie, a hamlet in Scotland. Geologists believe that the Rhynie fossils owe their superb preservation to periodic flooding from a nearby, silica-rich spring. The fossil plants, such as *Rhynia*, bear swollen roots, suggesting that fungi were already symbiotic with plant roots four hundred million years ago.

A form suggesting the most ancient members of the latest kingdom is still living today. This is *Psilotum nudum*, an obscure plant dwelling in greenhouses and in Florida, Pacific Islands, and other sunny climes. *Psilotum* is a system of stems, a mere bundle of green growing sticks. It releases spores into the air that produce sperm, which can swim in most soil films or across puddles. After fertilization the embryo gives rise to new shoot growth. Like a bryophyte, *Psilotum* lacks roots and seeds. But unlike a bryophyte, it has a vascular system and stands upright. Lacking leaves, photosynthesizing along its stems, it may resemble the earliest forms of plant life.

Modern mosses and liverworts also suggest the shapes of primitive greenery. These bryophytes overcame the algal dependency on a fluid surround by bringing water to land with them. The mere existence of leaves in mosses and liverworts gave them a big edge over the ancient psilophytes which, like *Psilotum*, were limited in their ability to gather light.

But, unlike vascular plants, the bryophytes never evolved structural support. Bryophytes to this day never grow more than a few centimeters high. They are vulnerable to the tricks of vascular plants that can easily overtop them, robbing them of sunlight.

Although no one is sure and the fossil evidence is scanty, many botanists believe that the simpler, more aquatic bryophytes evolved earlier than did the structurally more complex and dryness-resistant plants. Bryophytes are soft-bodied; their fossil record is decidedly poor. Modern bryophytes are utterly dependent on surface waters; they have no roots to scavenge for water down into the soil. But they are by no means fragile creatures limited to swamps, pond edges, river rocks, and waterfalls. Some live in areas of seasonal moisture, growing mainly during the wet season. Others, notably the ingenious sphagnum moss, are the sponges of the land. They are world-class water scavengers, capable of holding up to a thousand times their own body weight in water, storing it for dry times. A mound of sphagnum moss, moreover, employs the dead in the task of water retention. Only the surface of the mound is alive, but the moss corpses in the interior and lower reaches retain water for their descendants.

Most of plant diversification occurred in the vascular plants with tough bodies and conductive tissue. These organisms grew up, literally. Horsetails, for example, were among the first organisms to tower into the air. Modern horsetails, called "scouring rushes," have silica in their photosynthetic stems. They were used by European settlers of North America to scrub pots and pans. But these silica-stiffened organisms grew much larger in the past than do their descendants today. Ancient horsetails in primeval forests, in the Devonian period 410 million years ago, stretched up to fourteen meters high.

Primeval Trees

Rhynia-type plants seem to have evolved into many extant and now-extinct forms. The ancestral

71

Glossopterus scutum, fossil seed fern. One of the extinct groups of cycadofilicales whose trees made up the first forest over 225 million years ago before the evolution of the first dinosaurs. These ancient forests, crushed beneath the surface by floating tectonic plates, became coal.

vascular form probably gave rise to progymno-sperms—an extinct lineage that branched off in one direction to become tropical seed ferns, which themselves later gave rise to the flowering plants. Another branch became the conifers that brachiosaurs dined upon and that lived on to survive meteor impacts and ice ages. The primitive *Rhynia*-type vascular plant also diverged to become ginkgos, spore-releasing ferns, horsetails, and *Psilotum*. The branching talents of the original stem-maker thus came to enshrub and enforest the world.

A huge group of vascular plants, as important as the dinosaurs to the animal kingdom, have gone extinct. Known as the cycadofilicales or seed ferns, none of these trees—which looked like overgrown pineapples—are alive today. They were not ferns at all. Unlike modern ferns, they made conspicuously large seeds. These seeds (not directly related to modern seeds) were a major evolutionary innovation. Seeds can wait through a drought or cold spell. They can survive a lack of light. Seeds were as crucial to the dispersion of plants as water-tight eggs were for the great diversification of reptiles.

Possibly the first plants to produce seeds, the cycadofilicales abounded 345 to 225 million years ago, before any dinosaurs. They were the makers of the earliest forests. Leaves of the genus *Glossopteris* (Greek for "tongue-leaf") are common fossils in rocks deposited in the southern supercontinent of Gondwanaland. [PLATE 71] Exposed to powerful tectonic forces, Gondwanaland cracked and the pieces drifted apart on continental plates 200 million years ago. Those pieces are now called South America, Africa, Australia, India, and Antarctica. *Glossopteris*, like more than 99 percent of the plant and animal species recorded in the fossil record, is extinct.

Neither *Glossopteris* nor any of its once-successful relatives survived to inhabit the southern continents Gondwanaland has become. But, once upon a time, greening the world for over a hundred million years, forests of seed ferns swayed in warm winds from the southern reaches of Gondwanaland to the tropics of Laurasia, the northern supercontinent. Now, after 125 million years, the forests of Gondwanaland exist as a semipetrified, energy-rich plant refuse: coal.

By the end of the Devonian period and the start of the aptly named Carboniferous (Mississipian and Pennsylvanian) 360 million years ago, Earth was forested. Whether from Rhode Island, Edinburgh, or western Pennsylvania, coal of this age is replete with remains of leathery leaves, thick roots, and scaly bark. In the basement of the Biological Laboratories at Harvard University are "coal balls" that were hauled away from their sites of origin in Illinois and Kansas. Many are taller and, because they are spherical, stouter than a man. Chopping through them, or peeling off their surfaces with acid-treated acetate, reveals the source of ancient plant tissue: leaves, bark, roots, and flowerless sex organs hardly the worse for 290 million years of burial.

Measured by genera and higher taxa lost, the mass extinctions of the Permo-Triassic 245 million years ago were far more devastating than the better-known end-of-Cretaceous event that extinguished all the dinosaurs. A major factor in the Permo-Triassic extinctions may have been expansion of glaciers or a long period of profound cold—perhaps itself generated by a comet or meteor impact that darkened the skies with debris sent into orbit. Seed ferns were tropical plants. The seedlings of seed ferns and the trees themselves were vulnerable to bitter cold. Before all the seed ferns became extinct, however, at least one of their ancestors gave rise to plants that could withstand freezing temperatures— the conifers.

Conifer fossils are older than those of flowering plants. Fossil seeds of conifers are detectable as raised portions on the underside of female cone scales. Spruce, cedar, pine, and many other cone-bearing trees and shrubs alive today remain green all year long. So did many of their ancestors, adept at surviving arid, wintry conditions. The pollen of

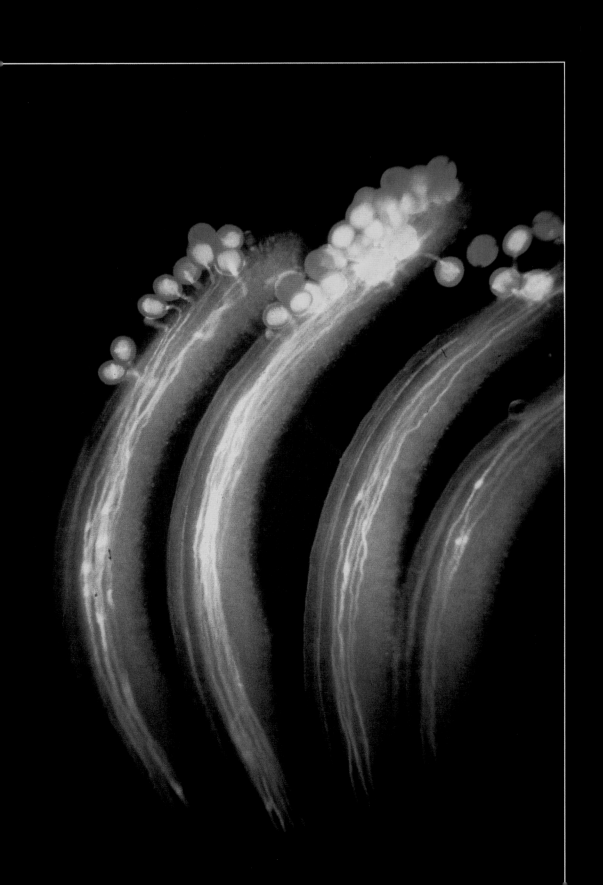

conifers is wind-borne. Fertilization of conifers leads to formation of seeds in the shelter of female cones. [PLATE 72] This change, from the ancestral method of releasing delicate water-borne sperm or short-lived spores such as those shed from the underside of fern fronds and destined to make tiny gametophyte plantlets, permitted evergreen conifers to dominate lands of seasonal ice and snow or aridity, as they do to this day.

Floral Persuasion

In contrast to all cone-bearing, naked-seed plants, the flowering plants have encased seeds—the result of the growth of the flower's ovaries into fruit. More than a quarter million species of flowering plants inhabit Earth. The seeds of these angiosperms bear embryos and the female organs transform to encase them.

Humans have a special relationship with angiosperms. Our primate ancestors lived among African flowering trees and fed, in part, on fruits—which had evolved lucious colors, arresting aromas, and other tempting qualities insofar as such qualities seduced us into becoming involved in their reproduction. Mammals dispersed the encased seeds and, by defecating, enriched the soil where angiosperms would sprout. Ancestors more closely resembling modern humans no longer lived in trees, but they kept their nimble hands and binocular vision while living in a new landscape that was another cross-kingdom creation. The grassland savannas were the workings of angiosperms that invented a way to grow from the base rather than their tips. Savannas were equally the creation of the large herbivores, whose grazing killed tip-growing forbs and young trees, thus "naturally selecting" the grasses. Finally, the savanna would have been impossible without its recyclers: protoctists and bacteria which digested cellulose in enlarged fore- and hind-guts coevolved in the large mammalian grazers.

Even today our species has a special relationship

Papaver somniferum, a poppy. Phylum: Angiospermophyta. Kingdom: Plantae. The name of the common poppy means "sleep-bringing" from the effects of its acrid and narcotic juice. Angiosperm plants, whose evolution dovetailed with that of mammals, still cast their floral spells on human eaters, drinkers and lovers. E. O. Wilson's biophilia theory suggests that we have genetically-embedded patterns of emotional response to other life forms. The color, smells and tastes emitted by flowering plants captivate with all the power of their 100-million-year aesthetic legacy.

with angiosperms. Angiosperm grains, fruits, stems, leaves, and roots are our primary foods—directly, or indirectly by way of domesticated livestock. (The only exceptions are intensive fishing communities.) We surround ourselves with furniture that is often made from the lignin of forest angiosperms. Angiosperms have taught us pleasure in nurturing them in flower gardens. Even the image of a branching tree to explain evolutionary phylogeny is easy to understand in part because of our ancient familiarity with the growth patterns of flowering trees.

Charles Darwin called the origin of the flower an "abominable mystery." [PLATE 73] Beautifully fossilized flowers and seeds indicate that flowering plants appeared in the middle latitudes of the northern hemisphere at least by 124 million years ago in the mid Cretaceous. Thus, within sixty million years of the last of the huge, flowerless, coal swamp trees, flowers had evolved and spread. Crab grass, philodendron, mother-in-law's tongue, spiderwort, Indian corn, pumpkins, tulips, coconut palms, and willow trees are all representative of flowering plants. Although the Amazon rain forest contains a concatenation of flowering plants, as little as 10,000 years ago that rain forest extended to only two percent of its present area. Like mammals, flowering plants—especially florid tropical jungles—are recent evolutionary phenomena.

Some plants have evolved such a high degree of interdependency with humans that they no longer survive in the wild. A graphic example of such intimacy is *Zea mays*, Indian corn. [PLATE 74] Evolved from teosinte, an obscure American grass, corn of many varieties now tower over people in season around the world. Simply by choosing seeds from the sweetest and most bountiful cobs year after year, people have rendered *Zea* entirely dependent. Unless removed and individually husked and planted by human hands or agricultural machinery, corn cannot reproduce. Absent this assistance, the kernels, trapped inside the fibrous covering, never sprout.

The "green revolution"—the vast increase in population due to development of agriculture and, subsequently, cities—is, from a biospheric perspective, a major success story for flowering plants. Like the *Atta* ants that tend their fungal gardens in rows, human ingenuity and resources—farm animals, fossil-fuel-driven tractors, fertilizers, irrigation, and biotechnological apparatus—have been devoted to maintaining the livelihood of our favorite plants. Our primate brains, evolving in a world of flowering plants, are still devoted to preserving and extending that verdant, nurturing world. Our attractions to

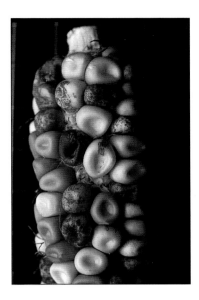

74
Zea mays, maize. Phylum: Angiospermophyta. Kingdom: Plantae. Colorful maize kernels on cob of Indian corn. Commercially produced corn requires husking by human hands or machines. This requirement is testimony to thousands of years of co-evolution of human beings and *Zea mays.* Today such corn can not reproduce without human agricultural intervention in its life process.

angiosperms are deep and instinctive—so much so that bottled essential oils are sold as perfumes, foods and drinks are artificially fruit-flavored, and clothes and toys are dyed shades of red, yellow, and orange—"warm" colors used in the original advertising campaign by which plants produced attractive and tasty incentives that enticed animals to do for them much of the work of fertilization.

Plants, too, have animals to disperse their offspring. We eat grapes but spit out the seeds, disseminating the plant. Bitter seeds and the hard pits of fruits would merit the adjective "clever" were they the fabrications of an organism with a hefty brain. Like colorful grocery packaging, bright and flavorful fruits with inedible or discardable cores manipulate the animal into collecting and spreading the offspring of the plant.

In an example of the growing intimacy among the many beings cohabiting the biosphere, these immobile, muscleless, and brainless beings—plants—have succeeded in appropriating the very powers of restlessness and active perception that separate them from beings to which they are commonly presumed to be inferior—animals. Like the symbiotic anastomosis of branches on the tree of life, the merging of plant reproduction with animal sensitivity and taste is a demonstration of life's

Satellite image of Earth from space showing the major zones of dominant vegetation: forest, desert, mountains, and other ecosystems. The continents here are seen as the raised portions of enormous moving plates that have changed their position over the history of the planet. Evolutionary biology and the new plate tectonic-based geology complement each other in broadening our view toward our living planet and its ancient history. A unified biosphere, Earth's surface is the chemically active aggregate of sunlight-transmuting, gas-exchanging, gene-trading, environmentally transformative life forms.

considerable powers of synergy and convergence. Living beings not only compete and struggle; they also associate and work together.

Solar Economy

We bipedal mammals like to think ourselves king of the Earthly hill, the most evolved form of life. But the argument might just as well be made in favor of flowering plants. They lack brains and speech — but, then, they don't need them. They borrow ours.

With our vaunted intelligence we have been Johnny Appleseeds, spreading fruit trees and grasses around Earth's surface. By tapping more directly than any previous animal into past and present photosynthetic powers, we raise the stakes of life on Earth. For, make no mistake about it, the solar economy has, with humans, entered a new phase.

Peter Vitousek, using satellite imagery, estimates that forty percent of the ice-free land surface of the globe is under agricultural cultivation; very little arable land remains untilled. [PLATE 75] Humanity annually uses the energy equivalent of 18 trillion kilograms of coal — about 3.6 metric tons for every man, woman, and child on the planet. This total energy is used, in part, to retrieve 327 billion kilo-

grams of iron, 90 billion kilograms of gypsum, and similarly staggering quantities of other materials. It is also used to generate and retrieve an estimated 540 billion kilograms of wheat, 90 billion kilograms of gypsum and 92 billion kilograms of seafood.

As fossil fuels and solar energy integrate into factory and machine production and into global husbandry and agriculture, more plants, animals, and microbes come to depend on the technological system now evolving. Nonrenewable resources are consumed, creating evolutionary innovations in the form of new biospheric waste: insecticides, polyvinylchloride, styrofoam, rayon, and latex paints. The gaseous byproducts of burning long-buried energy sources perturb or alter irreversibly the complex system of planetary physiology. Carbon dioxide accumulates in the atmosphere. Letting in visible light, but trapping reflected heat, this greenhouse gas may increase planetary temperatures — perhaps even melting polar ice, thus swamping coastal cities. Meanwhile, multiple extinctions follow from buzzsawing and bulldozing trees, killing some species directly but upsetting far more by destructive incursion into their living space. Nonetheless, the very energy our species uses to wreak habitat havoc comes ultimately from photosynthesis. For good or evil, novelty or status quo, nature is empowered by solar fire. The energy for violence also comes from plants.

Ever since *Homo sapiens* evolved, plants have fed, clothed, and sheltered us. From maternity ward vase to soft brown grave they accompany us on our biospheric journey. Borne on plants listing sunward, flowers symbolize peace, life, beauty, hope, femininity, and the sun.

Flowers, like tropical fish in an aquarium, elevate and calm. Fomenting biophilia, they are mental medicine, provoking our spirits. But flowering plants—all plants—are more than decorative. They are indispensable to an environment that can support humans. Their descendants will keep our descendants company. Spider plants, NASA reports, recycle trace pollutants in enclosed environments such as a space capsule. *Nymphaea*, a water lily, purifies drinking water. Long trips into space are inconceivable without plants to grow as food. Perhaps in seven generations your great, great, great, great, great grandaughter will look down at her toes—and see a wildflower poking up from a crack in the surface of Mars.

So, what is life?

Life is the transmutation of sunlight. It is the energy and matter of the sun become the green fire of photosynthesizing beings. It is the natural seductiveness of flowers. It is the warmth of the tiger stalking the jungle in the dead of night. Green fire converts wildly to the red and orange and yellow and purple sexual fire of flowering plants. Expanding, developing lignin, green beings raised up the biosphere and spread it horizontally. As fossils these beings trapped the original gold of the sun, stocking wealth only recently released in the human crucible of the solar economy. But the arrow in all these transformations must eventually become a loop that encloses the autopoietic exigencies of plants. We may be an intelligent life form but our very intelligence depends on that part of us we now recognize as photosynthetic. As life transmutes solar fire into all the energy and matter cycles of the biosphere, we pay homage to the ingenious ascension of the living plant. ■

9

Sentient Symphony

Owing to the imperfection of language, the offspring is termed a new animal; but is, in truth, a branch or elongation of the parent, since a part of the embryon animal is or was a part of the parent, and, therefore, in strict language, cannot be said to be entirely new at the time of its production, and, therefore, it may retain some of the habits of the parent system.

—**Erasmus Darwin,** *Zoonomia*

They say that habit is second nature.
Who knows but nature is only first habit?

—**Blaise Pascal,** *Pensées*

Thinking and being are one and the same.

—**Parmenides**

A Double Life

What is life? Two crucial traits are that life produces (autopoietically self-maintains) and reproduces itself. Then there is inherited change: DNA and chromosome mutation, symbiosis, and sexual fusion of growing life when combined with natural selection means evolutionary change. Nonetheless, autopoiesis, reproduction, and evolution only begin to encompass the fulness of life.

We have glimpsed ways of describing what life is: a material process that sifts and surfs over matter like a strange, slow wave; a planetary exuberance; a solar phenomenon—the astronomically local transmutation of Earth's air, water, and received sunlight into cells. Life can be seen as an intricate pattern of growth and death, dispatch and retrenchment, transformation and decay. Connected through Darwinian time to the first bacteria and through Vernadskian space to all citizens of the biosphere, life is a single, expanding network. Life is matter gone wild, capable of choosing its own direction in order to forestall indefinitely the inevitable moment of thermodynamic equilibrium—death. Life is also a question the universe poses to itself in the form of a human being.

Life is manifest on Earth as five kingdoms, each revealing from a different angle this mystery of mysteries. In a very real sense, life is bacteria and their progeny. Every available piece of real estate on this planet has become inhabited by subjects of the Kingdom Monera: by the enlightened producer, the tropical transformer, the polar explorer. Life is also the strange new fruit of individuals evolved by symbiosis. Different kinds of bacteria merged to make protoctists. When conspecific protoctists merged the result was meiotic sex. Programmed death evolved. Multicellular assemblages became animal, plant, and fungal individuals. Life is thus not all divergence and discord but also the coming

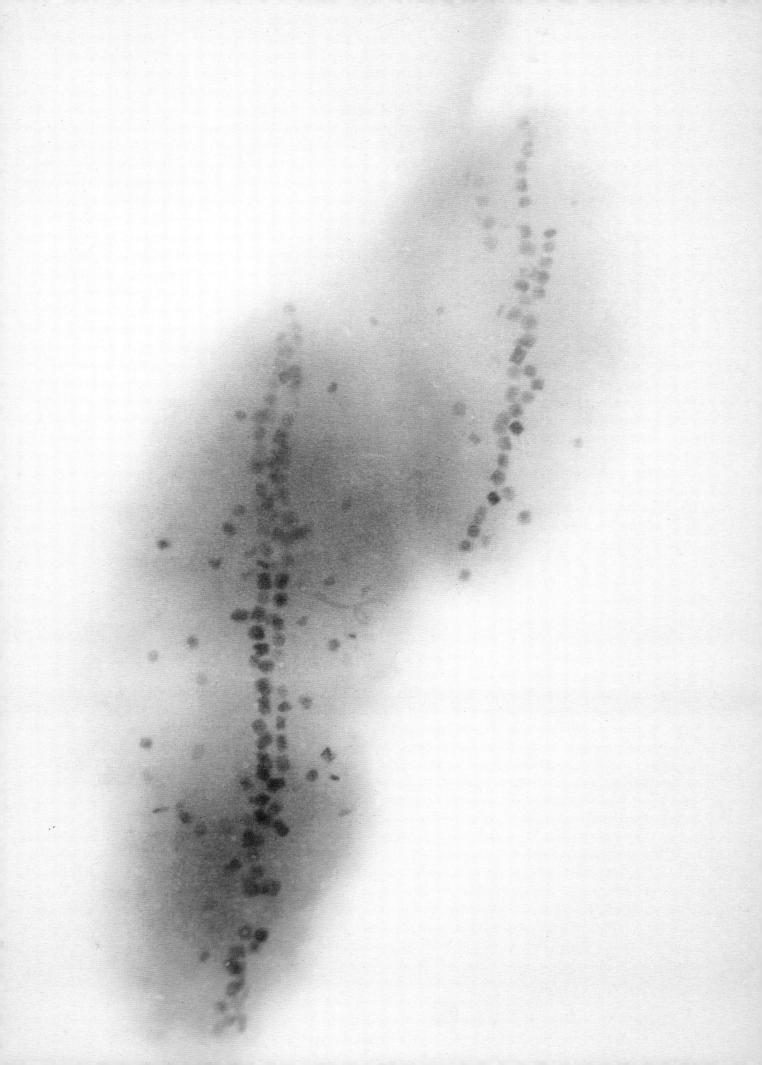

Humans, we are told, are special. We have upright posture (allowing us to think of ourselves as literally "above" other species). We have opposable thumbs (man the tool user), linguistic abilities (man the storyteller), a superanimalistic soul (Descartes' distinction). We have, at least in the western hemisphere, a cultural tradition of seeing ourselves as being in a position of moral stewardship over the rest of life. Even in the absence of God, we imagine ourselves to possess a unique capacity to destroy the planet (via nuclear weaponry) or to swiftly change atmosphere and climate.

Even such an ardent foe of the idea of progress in evolution as Stephen Jay Gould (and he is not alone) proposes that whereas humans can evolve quickly through "cultural selection," all other forms of life on Earth are shackled to the ancient, plodding system of "natural selection." But the sheer number of traits listed to explain human uniqueness is enough to arouse suspicion. Among the dazzling array of reasons implying our superiority over the rest of life, one scientific argument stands out to us in curious contrast to the rest: humans are the only beings capable of wholesale self-deception.

This claim is based on early humans' presumably delusionary belief in the afterlife. Before written history our ancestors buried their dead with food, weapons, and herbal medicines of little use to corpses. How ironic that we, in seeking examples of our superiority over the rest of life, have finally congratulated ourselves on a trait that threatens to negate all the others! Although members of other species trick one another, humans are the expert self-deceivers: as the best symbol users, the most intelligent species, and the only talkers, we are the only beings accomplished enough to fully fool ourselves.

Little Purposes

Freud's understanding of unconsciousness as repression—painful memories are pushed away from the conscious mind—has diverted attention from another way in which actions become unconscious. Not avoidance but extended focus can make an action automatic, second nature. A speech is "learned by heart." A practiced typist no longer glances at the keyboard. Virtually any activity when memorized subsides from conscious attention. The heart pumps, the kidneys filter in autonomic quasi-perfection. Over breathing and swallowing, normally automatic, the willful organism can exert some voluntary control and modulation.

Now here is a strange thought: Perhaps we mammals remain unconscious of inborn physiology because under pressure of survival our ancestors consciously practiced their skills to unconscious perfection. Although modern science does not yet offer us a mechanism that transmits the learned habits of one generation to the physiology of the next, experience shows that conscious can become unconscious with repeated action. The gulf between us and other organic beings is a matter of degree, not of kind. Taken together, the vast sentience comes from the piling up of little purposes, wants, and goals of uncounted trillions of autopoietic predecessors who exercised choices that influenced their evolution. If we grant our ancestors even a tiny fraction of the free will, consciousness, and culture we humans experience, the increase in complexity on Earth over the last several thousand million years becomes easier to explain: life is the product not only of blind physical forces but also of selection in the sense that organisms choose. All autopoietic beings have two lives, as the country song goes, the life we are given and the life we make.

Habits and Memory

Physicist Howard Pattee laments the simplistic misapplication of deterministic physics to biology. Pattee takes issue with biologists who are too quick to use classical physics to justify an understanding of life as a mechanical phenomenon. A general property distinguishing life from nonliving matter is its historical coherence, expressed as the potential to evolve. "Even though we do not understand the mechanism," Pattee continues, "the only conclusion I have been able to justify is that living matter has distinguished itself from nonliving matter by its ability to achieve greater reliability in its molecular hereditary storage and transmission processes than is obtainable in any thermodynamical or classical system."[5] The hints and hunches must be replaced by firm detail, yet the extraordinary storage and transmission processes of life for molecular heredity, mentioned by Pattee, may be robust enough to encompass the phenomenon postulated by Butler: phylogenetic "memorization," the conversion of the conscious strivings of one generation into the activities and, eventually, the physiologies of the next.

Although we fail as yet to see how an organism's or even a species' voluntary habits can become the physiology of a future generation via a material basis of heredity, we are fascinated by Butler's suggestion. We know, for example, that many organic beings acquire new heritable traits by symbiogenesis and that a vast array of others, not only people, are capable of learning. Ecosystems grow increasingly complex and sensitive; processes practiced in them repeatedly by one generation may become easier for the next. More open-minded investigation is needed. Objections may be leveled against Butler's ideas, yet he cannot be accused of the atavistic thinking which clings to humanity's separate status. Covertly considering ourselves divine, under the scientific rubric of "cultural evolution," or by dint of that other desperate euphemism, our "big brains," we are probably now more ecologically impoverished than we would be if, a century ago, we had adopted Butler's notion of all life as a conscious continuum.

Butler did not object to evolution but to the loss of the richness of the earlier, more lively views, in which living beings themselves were involved in natural selection:

> According to Messrs. Darwin and Wallace, we may have evolution, but are on no account to have it as mainly due to intelligent effort, guided by...sensations, perceptions, ideas. We are to set it down to the shuffling of cards....According to the older men, cards did indeed count for much, but play counted for more. They denied the teleology of the time—that is to say, the teleology that saw all adaptation to surroundings as part of a plan devised long ages since by a quasi-anthropomorphic being who schemed....This conception they found repugnant alike to intelligence and conscience, but, though they do not seem to have perceived it, they left the door open for a design more true and more demonstrable than that which they excluded.... They made the development of man from the amoeba part and parcel of the story that may be read, though on an infinitely smaller scale, in the development of our most powerful marine engines from the common kettle, or of our finest microscopes from the dew drop. The development of the steam-engine and the microscope is due to intelligence and design, which did indeed utilize chance suggestions, but which improved on these, and directed each step of their accumulation, though never foreseeing more than a step or two ahead, and often not so much as this.[6]

Existence's Celebration

For nineteenth century Englishmen of science it was natural and expedient to invoke Newtonian mechanics and conceive of life as Newton's matter: blind bits predictably responding to forces and natural laws. Like some piece of well-made clockwork, the world was donated or its mechanism manufactured by a transcendent god—a god that then stood outside its creation.

This was the new view of evolution: God, if it existed, was Newton's God. Not an active interloper in human affairs, it was the god of the mathematicians, the geometer god, who made the laws and then sat by and watched those laws play out. But an older view left room for a kind of god, too—a more active god. This was the view that Samuel Butler attempted to resuscitate—that life itself was godlike. There was no grand design, but millions of little purposes, each associated with a cell or organism in its habitat.

To the neo-Newtonians, the Darwinians, free will had been all but banished from the universe because the universe was portrayed as a mechanism and mechanisms do not have consciousness. For Descartes, God continued to have consciousness and people did insofar as they were in touch with God. But when Darwin showed through painstaking work that people too could be explained by the mechanism of natural selection, consciousness suddenly became redundant in the human world as well. Butler brought consciousness back in by claiming that, together, so much free will, so much behavior becoming habit, so much engagement of matter in the processes of life, so many decisions of where, how, and with what or whom to live, had shaped life, over eons producing visible organisms, including the colonies of cells called human. Power and sentience propagate as organisms. Butler's god is imperfect, dispersed.

We find Butler's view—which rejects any single, universal architect—appealing. Life is too shoddy a production, both physically and morally, to have been designed by a flawless Master. And yet life is more impressive and less predictable than any "thing" whose nature can be accounted for solely by "forces" acting deterministically. The godlike qualities of life on Earth include neither omniscience nor omnipotence, although an argument could be made for earthly omnipresence.

Life, in the form of myriad cells, from luminescent bacterium to lily-hopping frog, is virtually everywhere on the third planet. All life is connected through Darwinian time and Vernadskian space. Evolution places us all in the stark but fascinating context of the cosmos. Although something odd may lurk behind and before this cosmos, its existence is impossible to prove. The cosmos, more dazzling than any sect's god, is enough. Life is existence's celebration.

Butler's forgotten theory intrigues us. The mind and the body are not separate but part of the unified process of life. Life, sensitive from the onset, is capable of thinking. The "thoughts" both vague and clear, are physical, in our bodies' cells and those of other animals.

In comprehending these sentences, certain ink squiggles trigger associations, the electrochemical connections of the brain cells. Glucose is chemically altered by reaction of its components with oxygen, and its breakdown products, water and carbon dioxide, enter tiny blood vessels. Sodium and calcium ions, pumped out, traffic across a neuron's membranes. As you remember, nerve cells bolster their connections, new cell adhesion proteins form, and heat dissipates. Thought, like life, is matter and energy in flux; the body is its "other side." Thinking and being are the same thing.

If one accepts this fundamental continuity between body and mind, thought loses any essential difference from other physiology and behavior. Thinking, like excreting and ingesting, results from lively interactions of a being's chemistry. Organism thinking is an emergent property of cell hunger, movement, growth, association, programmed death, and satisfaction. Restrained but healthy former microbes find alliances to make and behaviors to practice. If what is called thinking results from such cell interactions, then perhaps communicating organisms, each of themselves thinking, can lead to a process greater than individual thought. This may have been what Vladimir Vernadsky meant by the noosphere.

Gerald Edelman and William Calvin, both neuroscientists, have each proffered a kind of "neural darwinism." Our brains, they say, become minds as they develop by rules of natural selection.[7] That idea may provide a physiological basis for Butler's insights. In the developing brain of a mammalian fetus, some 10^{12} neurons each become connected with one another in 10^4 ways. These cell-to-cell adhesions at the surface membranes of nerve cells are called synaptic densities. As brains mature over ninety percent of the cells die! By programmed death and predictable protein synthesis connections selectively atrophy or hypertrophy. Neural selection against possibilities, always dynamic, leads to choice and learning, as remaining neuron interactions strengthen. Cell adhesion molecules synthesize and some new synaptic densities form and strengthen as nerve cells selectively adhere and as practice turns to habit. Selection is against most nerve cells and their connections but it is nevertheless for a precious few of them. Of course, more work is needed before the physical basis of thought and imagination can be understood, but selective death in a vast field of proliferating biochemical possibilities may apply to minds as it does to evolution.

The peculiarly curved early embryos of birds, alligators, pigs, and humans are remarkably similar. Developing from a fertile egg, all display a stage with gill slits—whether the hatched or born animal breathes oxygen from air or water. The slits that close behind the ears in the human fetus attest to our common ancestry with fish, whose gill slits function in the adult. Human embryos have tails. Living matter "remembers" and repeats its origins to arrive in the present. In a Butlerian world, the materials of living beings are molded by life, over and over, for millions of generations. Creating a sense of *déjà vu*, the embryo represents a once-unconscious process, now again—at a different level—brought to consciousness.

Superhumanity

A transhuman being, superhumanity, is appearing, becoming part of the sentient symphony. It is composed not only of people but of material transport systems, energy transport systems, information transport systems, global markets, scientific instruments. Superhumanity ingests not only food but also coal, oil, iron, silicon.

The global network that builds and maintains cities, roadways, and fiber-optic cables grows by leaps and bounds. In Nigeria, for example, the population is expected to reach 216 million by the year 2010, double that of 1988. Unchecked, such growth would bring the number of Nigerians to more than 10,000 million by the year 2110—twice the present global human population. Our stupendous population taps a significant proportion of the solar energy reaching Earth's surface. The raw energy of photosynthesis, past and present, and transformed into edible plants, animal fodder, geological reserves, and human muscle and brain, supports the massive construction of the transcontinental urban ecosystem and even—"biting the hand that feeds it"—the razing of forests capturing and converting solar energies. [PLATE 77] As the system expands using genetic and atomic technology, its operations become more elegant and cohesive. The potential for disaster also increases.

Superhumanity is neither a simple collection of humans nor something other than aggregated humans and their devices. Plumbing, tunnels, water pipes, electric wires, vents, gas, air conditioning ducts, elevator shafts, telephone wires, fiber-optic cables, and other links enclose humans in a rapidly growing net. The way superhumanity behaves is in part the result of uncountable and unaccountable economic decisions made by people—singly and in groups—within the context of an increasingly planetary capitalism. "The problem with money," says a character in a recent film, "is that it makes us do things we don't want to do."

Whether or not superhumanity's tendencies are conscious beyond us, individual humans should not be surprised if the aggregate of planetary humanity shows unexpected, emergent, seemingly purposeful behaviors. If brainless bacteria merged into fused protists, that cloned and changed themselves over evolutionary time into civilization, what spectacle will emerge from human beings in global aggregation? To deny the existence of superhumanity by insisting it is merely the sum of human actions is like claiming that a person is merely the sum of the microbes and cells that constitute the body.[8]

Expanding Life

Life today is an autopoietic, photosynthetic phenomenon, planetary in scale. A chemical transmutation of sunlight, it exuberantly tries to spread, to outgrow itself. Yet by reproducing, it maintains itself and its past even as it grows. Life transforms

77
Biology merging with technology. The trans-Alaskan pipeline, snaking across the countryside from Prudhoue Bay to the port of Valdez in Alaska. Graphic example of the "plumbing" of the external oil-using, circulatory system of "superhumanity"—globally connected, industrially contoured, goods and services trading, telecommunicating humanity taken as a global whole. Vladimir Vernadsky in his speculation on the biosphere posited the development of a global layer of thinking and technology called the noosphere. This photograph represents one of the developing conduits of Vernadsky's noosphere.

to meet the contingencies of its changing environment and in doing so changes that environment. By degrees the environment becomes absorbed into the processes of life, becomes less a static, inanimate backdrop and more and more like a house, nest, or shell—that is, an involved, constructed part of an organic being. The members of thirty million species interacting at Earth's surface continue to change the world.

Coming to understand life afresh, we find that species of organisms diverge into new kinds, yet earlier patterns never entirely disappear. Old life forms, the bacteria that run the planet's ecology, are supplemented but not replaced. Although every distinct variety of nucleated being—every species of plant, animal, protoctist, and fungus—perishes, similar new taxa evolve from them or from their kind. Meanwhile, the underlying bacteria march symbiogenetically on.

We find that nature is not always "cutthroat," or, as the poet Alfred Tennyson put it, "red in tooth and claw." Living beings are amoral and opportunistic, as befits their needs for water, carbon, hydrogen, and the rest. They are fractally repeating structures of matter, energy, and information, with a very long history. But they are no more inherently blood-thirsty, competitive, and carnivorous than they are peaceful, cooperative, and languid. Lord Tennyson might just as well have cast nature as "green in stem and leaf."

Among the most successful—that is, abundant— living beings on the planet are ones that have teamed up. Moving inside (or perhaps forcibly dragged inside) another cell, the cyanobacteria that became chloroplasts in protoctist and plant cells weren't lost; they were transformed. So too with the mitochondria—once respiring bacteria—that give your finger muscles the energy to turn this page. Former bacteria, as themselves or parts of larger cells, are still the most abundant forms of life on the planet.

The strength of symbiosis as an evolutionary force undermines the prevalent notion of individuality as something fixed, something secure and sacred. A human being in particular is not single, but a composite. Each of us provides a fine environment for bacteria, fungi, roundworms, mites, and others that live in and on us. Our gut is packed with enteric bacteria and yeasts that manufacture vitamins for us and help metabolize our food. The pushy microbes of our gums resemble department store customers before a holiday. Our mitochondria-laden cells evolved from a merger of fermenting and respiring bacteria. Perhaps spirochetes, symbiotically faded to the edge of detectability, continue to squirm as the undulipodia of our fallopian tubes or sperm tails. Their remnants may move in subtle ways as our microtubule-packed brains grow. "Our" bodies are actually joint property of the descendants of diverse ancestors.

Individuality is not stuck at any one level, be it that of our own species or pondwater *Amoeba proteus*. Most of our dry weight is bacteria, yet as citizens swarming in crowded streets and office buildings, viewing television, traveling in cars, and communicating by cellular and facsimile phone, humans disappear in a global swirl of activity, overwhelmed by emergent structures and abilities that could never be accomplished by individuals or even tribes of human predecessors. No single human can speak to another human, in real time, thousands of miles away. No single human could stand on the moon. These are emergent abilities of superhumanity. Our global activities bring to mind the social insects, except that our "hive" is nearly the entire biosphere. [PLATE 78]

Inextricably embedded in the biosphere, this superhuman society is not independent. At its

78

Queen honey bee surrounded by female workers licking and touching her with their antennae. They are attracted by a chemical called a pheromone which they will distribute throughout the nest subsequently in a buzz of relative hyperactivity. Social organization acts at many levels and magnitudes in stages throughout nature's holarchy. Organisms become societies become superorganisms become individuals super-ordinating into societies again at higher scales.

greatest extent life on Earth—fauna, flora, and microbiota—is a single, gas-entrenched, ocean-connected planetary system, the largest organic being in the solar system. The upper mantle, crust, hydrosphere, and atmosphere of Earth remain in an organized state very different from that on the surfaces of our neighboring planets. Photosynthesis, respiration, fermentation, biomineralization, population expansion, seed germination, stampedes, bird migration, mining, transportation, and industry move and alter matter on a global scale. Life dramatically impacts the environment by producing and storing skeletons and shells of calcium phosphate and carbonate, by caching plant remains as coal and algae residues as oil. Great layers of minerals—sulfides of iron, lead, zinc, silver, and gold—remain in place where they were precipitated by hydrogen gas-producing bacteria.

Minerals not normally associated with life—aragonite, barite, calcite, francolite, fluorite—are produced as crystals and skeletons inside, and exoskeletons or shells outside, living organisms. Plants and microbes induce the formation of "inanimate" substances such as nickel, iron oxides, galena (lead sulfide), and pyrite (iron sulfide, or fool's gold). Humanity's cultures are ranked from a stone age through an iron to a bronze age. Some argue that with the advent of computers Earth has entered a "silicon age." But metallurgy preceded us: human metalworking followed the bacterial use of magnetite for internal compasses by 3,000 million years. *Pedomicrobium*, a soil bacterium found fossilized in gold samples, is thought to precipitate gold ions and thus accumulate gold particles in its sheath. Next to millions of cubic kilometers of tropical reefs built by coral and entire cliffs of chalk precipitated by foraminifera and coccolithophorids, human technology does not seem uniquely grand.

Our destiny is joined to that of other species.

When our lives touch those of different kingdoms—flowering and fruiting plants, recycling and sometimes hallucinogenic fungi, livestock and pet animals, healthful and weather-changing microbes—we most feel what it means to be alive. Survival seems always to require more networking, more interaction with members of other species, which integrates us further into global physiology. Despite the apocalyptic tone of some environmentalists, our species is on its way to becoming better integrated into global functioning. The same technology that now poisons other organisms, stultifies their growth, and does the same to us, may usher in the next major change in biospheric organization.

As anaerobic microbes teamed up to make the swimming ancestors of protist colonists, as mastigotes ingested but did not digest the mitochondria that made them able to invade oxygen-rich niches at Earth's surface, or as fungi and algae combined into lichens that colonized bare rock of dry land, so too the transport of life to new planetary bodies will require teamwork. Astronautics, computers, genetics, biospherics, telecommunications, and other forms of photosynthesis-based, human-sired technology will have to combine with the predecessor technology of other planetmates. The ultimate explosion of life onto its next frontier—that of space—will require the new technology of life itself. The vivification and terraformation, the coming to life of other planetary bodies is not only a human process. Some day recycling ecosystems inside spacecraft may feed humans voyaging to other planets. If humans are to reside in space or voyage beyond Earth's orbit, the plants that feed them, the bacteria that digest for them, the fungi and other microbes that recycle their wastes, and the technology that supports them will surely be along for the ride. The extension of our local thermodynamic disequilibrium into space necessarily will involve representatives of all five kingdoms that have forged new ecosystems, able to transfer energy and cycle matter in isolation from the mother planet, the original functioning biosphere of Earth.

The distinction between space colonization by machines alone or by life with machines mirrors a New Zealand newspaper debate Samuel Butler had with himself. Beginning in 1862 Butler contributed an anonymous article to *The Press* of Christchurch, New Zealand. At the time, he was sheep farming in the Upper Rangitata district of Canterbury Province. Entitled "Darwin on the Origin of Species, a Dialogue," the unsigned article generated spirited protest. Butler joined in, criticizing himself as well as others. Signing his divided opinions under different names, Butler ultimately argued for two diametrically opposed interpretations of machines.

In "Darwin among the Machines," a letter signed by one Cellarius and appearing in *The Press* on 13 June 1863, Butler held that machines were the latest form of life on Earth, posed to take over and enslave their human masters; the rate of evolution and reproduction of machines was prodigious, and without "war to the death" at once it would be too late to resist their world dominion. Then, in a 29 July 1865 article entitled "Lucubratio Ebria," Butler countered Cellarius by saying that human beings were not even human without clothes, tools, and other mechanical accessories. Machines were not a threat to human life, but its indissociable natural extension.[9]

If space vehicles do cut free of human influence, voyaging starry skies as they reproduce on their own, then Cellarius and other Luddites will be vindicated. If, however, machines in space flourish not alone but as intelligent enclosures for a wide variety of other life forms, then the author of "Lucubratio Ebria" will be proven correct.

We place our bets on the latter. Machines, we believe, will flourish in a tightly meshed interface with life—not only human life but a rich assortment of starlight-using life forms. People are essential for making the export of life into the night fantastic even possible. But, like the sperm tail which breaks off once the genetic message enters the egg, so human beings are ultimately expendable. Even without us, a hundred million more years of sun-driven planetary exuberance should be enough to get life off the Earth, star-bound. Other technological species might evolve. Besides, not only humans have begun space exploration.

Setting foot on the moon, Neil Armstrong proclaimed, "One small step for man, one giant leap for mankind." True, in a sense; but he overlooked vast numbers of bacteria on his skin and in his intestine that stepped with him. Life has been expansionist from the beginning. Once it gets a firm toehold in space it may kick off its human shoes and run wild.

Rhythms and Cycles

We and many other animals sleep and wake in cycles that repeat every twenty-four hours. Some ocean protists, dinomastigotes, luminesce when dusk comes, ceasing two hours later. So hooked are they into the cosmic rhythm of Earth that even back in the laboratory, away from the sea, they know the sun has set. Many similar examples abound because living matter is not an island but part of the cosmic matter around it, dancing to the beat of the universe.

Life is a material phenomenon so finely tuned and nuanced to its cosmic domicile that the relatively minor shift of angle and temperature change as the tilted Earth moves in its course around the sun is enough to alter life's mood, to bring on or silence the song of bird, bullfrog, cricket, and cicada. But the steady background beat of Earth turning and orbiting in its cosmic environment provides more than a metronome for daily and seasonal lives.

Larger rhythms, more difficult to discern, can also be heard.

Many types of life form encapsulating structures that protect them from temporary dangers of the environment. Propagules of a wide variety of types are miniaturized, viable representatives of mature organic beings. They range from bacterial and fungal spores to protoctist cysts; from plant spores, pollen, seeds, and fruit to the dry eggs of some crustaceans, insects, and reptiles. As such propagules proliferate, natural selection deals with them severely: many die or fail to grow.

Desiccation- and radiation-resistant, most propagules metabolize at an exceedingly slow rate. Spores of bacteria may lay in wait for a hundred years—until rain comes, or phosphorus abounds, or conditions otherwise become less dry and more permissive. Without any dormant seeds or resistant spores, humans survive extraordinary environmental hazards. Houses, clothes, railroads, and automobiles have made possible our expansion from the subtropical home to colder climes. Analogous to spores, cysts, and seeds, these protective structures protect us from harsh conditions. [PLATES 79 and 80]

Recycling greenhouses are enclosed dwellings that contain representative collections of Earth's life. They detoxify poisons and transform wastes into food and back again. One, designed by Santiago Calatrava, will span the entire roof of the Cathedral of Saint John the Divine in New York City. Such "artificial biospheres" miniaturize crucial processes of the autopoiesis of the global ecosystem.

The global ecosystem is not an ordinary organic being. The global system, like all living beings, is energetically open: solar radiation comes steadily in, dissipated heat moves steadily out. But unlike other beings, the global system is closed to material exchange. Apart from the occasional incoming meteor or comet, nothing enters. Apart from the

79 and 80
Progagules of *Adansonia*, baobob trees from the island of Madagascar. Their huge water-borne propagules are seeds. The appearance and evolution of durable desiccation resistant structures appears "fractally" at many levels from bacterial spores to protoctist cysts to the nuts of fruit trees to closed ecological systems proximately built by human beings but ultimately strange new fruit of the biosphere.

occasionally stalled geologic churning here and there of sediments into new crust and cooked gases, nothing leaves. All the matter used by life is recycled matter—reappearing matter that is never consumed.

No living cell or organism feeds on its own waste. Thus artificial ecosystems have biological importance that exceeds architecture or other human concerns. For the first time in evolutionary history, the biosphere has been reproduced, or better, has begun reproducing itself through humankind and technology. The generation of new "buds," materially closed living systems, within the "mother biosphere" resembles the structure of a fractal.

From a "green" or "deep ecology" perspective, humans do not dominate but are deeply embedded within nature. Artificial biospheres are the first buds of a planetary organic being, as "man-made" biospheres have the potential to duplicate biospheric, light-dependent self-sufficiency. People in NASA, the European Space Agencies, governments, private industry, academia and elsewhere are pondering these desiccation-resistant structures that sequester, like new Noah's Arks, samples of life—not in museums but in a living, self-sustaining form. The largest of the recycling structures is Biosphere 2 in the Sonoran desert at Oracle, Arizona. Ultimately, closed ecosystems are not artificial at all, but part of the natural processes of self-maintenance, reproduction, and evolution in a heat-dissipating universe.

For an organic being to survive in space, food must be replenished and waste disposal systems must work. Photosynthesis has stored solar energy in rocks as reserves of kerogen, oil, gas, iron sulfide, coal, and other substances. The planet's prodigal species now expends these reserves, using the energy to spread its populations. *Homo sapiens* spends the riches of eons; meanwhile, the rhythms of Earth, building up for ages, approach a crescendo. But

nature has not ended, nor does the planet require saving. The technological dissonance marks no end but a lull, a gathering of forces.

Global life is a system richer than any of its components. We alone among the animals build large telescopes and mine Archean diamonds. But, while our position cannot soon or easily be replaced, we are not in charge. The diamonds are made of carbon, a main element of life since its inception some 4,000 million years ago; and telescopes are lenses, parts of the compound eye of a metahuman being that is itself an organ of the biosphere.

The continuous metamorphosis of the planet is the cumulative result of its multifarious beings. Humankind does not conduct the sentient symphony: with, or without us, life will go on. But behind the disconcerting tumult of the present movement one can hear, like medieval troubadours climbing a distant hill, a new pastorale. The melody promises a second nature of technology and life together spreading Earth's multispecies propagules to other planets and the stars beyond. From a green perspective a keen interest in high tech and the altered global environment make perfect sense. It is high noon for humanity. Earth is going to seed. ■

Epilogue

Epilogue

There is not so much Life as talk of Life, as a general thing. Had we the first intimation of the Definition of Life, the calmest of us would be Lunatics!

—Emily Dickinson
(letter to Mrs. Holland, c. 1881)

Half a century ago the preeminent physicist Erwin Schrödinger, a humane and thoughtful scholar, approached, as science, the question of what life is. Working prior to the discovery of DNA and before knowledge accumulated on how enzyme proteins and chemical movement become the metabolism of life, Schrödinger nonetheless inspired the very search that led to a materialistic explanation of living processes. We are fortunate to engage Schrödinger's tradition after fifty additional years of scientific inquiry.

To function, the biosphere requires microbial diversity; to feel whole and at home most of us crave nature's variety. Perhaps we now are more alarmed about the human prospect than was Schrödinger because we live on a more populated Earth. Humans today clearly are threatened by the extinction, even before science can describe them, of so many of our planetmates. Plastics spread, tropical rain forests die out, coral reefs collapse. We wonder whether the growing understanding of life's autopoietic tendencies for expansion and control and the evolutionary history of planetary change in the wake of rapidly spreading life forms will make individuals less likely to buy packaged plastic goods, travel using fossil fuels, eat meat, or engage in other environmentally destructive activities. We doubt it!

Coming to know the varieties of life on Earth—life which, from pond scum to tigress, is connected to us through time and space—can serve to inspire. That excess is natural but dangerous we learn from the photosynthetic ancestors of plants. That movement and sensation are thrilling we experience as animals. That water means life and its lack spells tragedy we garner from fungi. That genes are pooled we learn from bacteria. Modern versions of our ancient aquatic ancestors, the protoctists, display versions of the urge to couple, and of our capacity to make choices. Humans are not special and independent but part of a continuum of life encircling and embracing the globe.

Sources of Illustrations

[**Plates**] **1.** NASA **2.** CNRI / Science Photo Library **3.** UPI / Bettman Archives **4.** Peter F. Allport **5.** Prof. Woodruff T. Sullivan III, University of Washington, Seattle **6.** NASA **7.** Dr. Johannes H.P. Hackstein, University of Nijmegen, The Netherlands **8.** Prof. Mary Beth Saffo, Arizona State University, Tempe **9.** Nancy Sefton/Science Photo Library **10.** Prof. Bernard Vandermeersch, Université de Bordeaux I **11.** NASA **12.** NASA/Science Photo Library **13.** Prof. Peter Westbroek, University of Leiden, The Netherlands **14.** Dr. Patrick Holligan, Marine Biological Association, Plymouth, England **15.** Prof. David Deamer, University of California, Santa Cruz **16.** NASA **17.** Prof. James A. Shapiro, University of Chicago **18.** Prof. Arthur Winfree, University of Arizona, Phoenix **19.** Prof. Oscar Miller/Science Photo Library **20.** Hans Reichenbach, Braunschweig, Germany **21.** Dr. L. Card/Science Photo Library **22.** Prof. Euan Nisbet, Royal Holloway and Bedford New College, University of London **23.** Prof. Dennis Searcy, University of Massachusetts **24.** Prof. Norbert Pfennig, University Konstanz, Germany **25.** Prof. Andrew Knoll, Harvard University **26.** Prof. Stjepko Golubic, Boston University **27.** Prof. Lynn Margulis, University of Massachusetts **28.** Elso S. Barghoorn (deceased) **29.** (top) Prof. Stanley Awramik, University of California, Santa Barbara; (bottom) Prof. Peter Westbroek **30.** Christie Lyons **31.** Prof. Tore Lindholm, Åbo Akademi, Finland **32.** David Chase (deceased) **33. - 34.** Christie Lyons **35.** Prof. Lynn Margulis **36.** Dr. Heinz Stolp, Bayreuth, Germany **37.** François Gohier/Science Photo Library **38.** Brian Duval, University of Massachusetts **39.** Walker/Photo Researchers **40.** Walker/Science Photo Library **41.** Prof. David John, Oral Roberts Medical School **42.** Prof. Frank E. Round, University of Bristol, England **43.** Christie Lyons **44.** Prof. Charles Cuttress **45.** Prof. Vivian Budnik, University of Massachusetts **46.** EM by Prof. Richard Linck, University of Minnesota; false color by David Gray, Woods Hole Oceanographic Institution **47. - 48.** Prof. Gonzalo Vidal, Uppsala University **49. -50.** Prof. J. Woodland Hastings, Harvard University **51.** R.O. Schuster **52.** Christie Lyons **53.** Dr. Johan Bruhn, University of Missouri **54.** Jukka Vauras **55.** William Ormerod **56.** James G. Schaadt **57. - 58.** Prof. Vernon Ahmadjian, Clark University **59.** Prof. Reinhard Agerer, Universität München, Germany **60.** Prof. Lynn Margulis **61.** Prof. Barbara Thorne, University of Maryland **62. - 64.** Anabel Lopez, Autonomous University of Madrid **65.** Biophoto Assoc./Photo Researchers **66.** Connie Barlow **67.** Prof. Jeremy Pickett Heaps, University of Melbourne, Australia **68.** Michael Dolan, University of Massachusetts **69.** Christie Lyons **70.** Brian Duval, University of Massachusetts **71.** Jim Frazier, from *The Flowering of Gondwana* by Mary White **72.** Prof. David Mulcahy, University of Massachusetts **73.** William Ormerod **74.** Prof. James A. Shapiro, University of Chicago **75.** NASA **76.** Prof. John F. Stolz, Duquesne University, Pittsburgh **77.** Bryan and Cherry Alexander **78.** Kenneth Lorenzen **79. - 80.** Prof. David Baum, Harvard University [**Charts**] **A.** Jeremy Sagan, Cornell University **B.** Charles Keeling, University of Hawaii, and Lloyd Simpson, Center for the Study of the Environment, Santa Barbara **C.** Prof. Lynn Margulis

Index

Book Design

Although there is no direct borrowing, I have been inspired by early scientific books, particularly those printed by Erhard Ratdolt (see Euclid's *Geometriae elementa*, 1482). I have tried to combine the typographic clarity and beauty of these works with a more contemporary sensibility. The text typeface, Minion, designed in 1990 by Robert Slimbach for Adobe Corporation, is loosely based on old style roman typefaces from the late fifteenth and sixteenth centuries. Likewise, the display typeface, Syntax, designed in 1968 by Hans Eduard Meier, is humanistic in structure.

– José Conde / Studio Pepin / Tokyo, Japan

■

Illustrations by Christie Lyons

Created and Produced by
Nevraumont Publishing Co., New York.
President, Ann J. Perrini